THE CONTROL OF INJURIOUS ANIMALS

BIOLOGICAL SCIENCE TEXTS

General Editor
Don R. Arthur, M.Sc., Ph.D., D.Sc., F.I.Biol.
Professor of Zoology,
King's College, University of London

SURVIVAL: MAN AND HIS ENVIRONMENT
DON R. ARTHUR, M.Sc., Ph.D., D.Sc., F.I.Biol.

THE INSECTS: STRUCTURE AND FUNCTION
R. F. CHAPMAN, B.Sc., Ph.D.

PROBLEMS IN QUANTITATIVE BIOLOGY
J. F. EGGLESTON, B.Sc., M.I.Biol.

STATISTICAL METHODS IN BIOLOGY
N. R. J. BAILEY, M.A., D.Sc.

AN INTRODUCTION TO GENETICS
C. M. M. BEGG, B.Sc., M.A., Ph.D., F.R.S.E.

AN INTRODUCTION TO BIOCHEMISTRY
E. O'F. WALSH, B.Sc., Ph.D., F.P.S., F.R.I.C.

AN INTRODUCTION TO ANIMAL PARASITOLOGY
J. D. SMYTH, M.A., Sc.D.

THE ECOLOGY OF ROCKY SHORES
J. R. LEWIS, B.Sc., Ph.D.

LIFE IN MUD AND SAND
S. K. ELTRINGHAM, B.Sc., Ph.D., M.I.Biol.

THE CONTROL OF INJURIOUS ANIMALS

J. M. Cherrett B.Sc., Ph.D., F.R.E.S.
J. B. Ford B.Sc., Ph.D., F.R.E.S., M.I.Biol.
I. V. Herbert B.Sc., M.S., Ph.D., M.I.Biol.
A. J. Probert B.Sc., Ph.D.

Department of Applied Zoology
University College of North Wales, Bangor

The English Universities Press Limited

ACKNOWLEDGEMENTS

We are grateful to Mr. J. Hobart and Professor Don. R. Arthur for their helpful comments on the entire text, and to Professor J. L. Harper and Mr. R. W. Howarth for their comments on certain sections.

ISBN 0 340 05225 2 Boards
ISBN 0 340 12365 6 Unibook

First printed 1971

The English Universities Press Ltd.
St Paul's House, Warwick Lane, London EC4P 4AH

Printed and bound in Great Britain by
Hazell Watson & Viney Ltd, Aylesbury, Bucks

EDITOR'S FOREWORD

In recent years there has been a new and enlightened attitude towards research and the teaching of biology in universities and schools, which has necessitated crossing the boundaries of such conventional subjects as Zoology, Botany, Mathematics, Physics and Chemistry. For only by such cross-fertilisation of disciplines can the subjects of biological reaction and interaction be fully comprehended. This book by Drs. Cherrett, Ford, Herbert and Probert is an example of the practical application of these ideas in an important economic field. At the present time there is conflict between those engaged in increasing the gross national productivity, those involved in the multi-ramifications arising from ever increasing world populations and those interested in conservation. Much has been written in one way or another on the polluted environment, and this, on a global scale, has implicated the use of pesticides, some of which have now been withdrawn from the market. Nevertheless, recognition of the fact that the populations of organisms can develop to the level of pests with deleterious effects on animal and human health and of reducing crop yields justifies the use of control measures. The authors of this book have attempted to define the pest concept, the meaning of pest control and, by rigorous examination, to discuss the methods employed. In this they have not shrunk from considering such issues as the side effects of control nor the economic implications. Their book presents a balanced, sane and informative account of problems of considerable significance in today's world.

It is intended for students interested in control technology, but is also essential reading for those with a practical bias in the broad field of zoology and in veterinary and human medicine.

Don. R. Arthur

PREFACE

In Chapter 1 of this book a pest animal is described as any animal which is noxious, destructive or troublesome to man, or to his interests. On the basis of this broad definition we have attempted to provide students who are developing an interest in control technology or in applied zoology in its broadest sense (Parasitology, Entomology, Veterinary and Human Medicine) with an introduction to present day ideas and techniques involved in the control of pest animals. In the Introduction and Chapter 1 the magnitude of modern pest problems and the underlying causes of pest outbreaks are discussed, together with the ecological basis for successful control. Chapter 2 deals with the preventive measures used to avoid pest damage and Chapters 3, 4, 5, 6, and 7 concentrate on the destruction of animals whose populations have grown to the point at which the animal has acquired economic pest status. A good deal of public concern has resulted from the publicity given to the side effects associated with current control operations, particularly those involving the use of poisonous chemicals, and these aspects of control are reviewed in Chapter 8. Finally, in Chapter 9, some of the social and economic considerations which have emerged as pest control has grown into an enormous worldwide undertaking are discussed.

In this latter half of the twentieth century the control of pest animals has no small part to play in the improvement and maintenance of the quality of human existence. More and more specialists from a diversity of scientific disciplines are being drawn into work on various aspects of this problem and inevitably the subject is developing its own language, with a vocabulary drawn from these various scientific specialities. We have assumed a basic knowledge of zoology and organic chemistry on the part of the reader but nevertheless some of the animals quoted as examples, and some of the technical terms which cannot be avoided, will be unfamiliar, and a glossary of these words is provided at the end of the book.

CONTENTS

INTRODUCTION

Man suffers very widely from the depredations of other animal species, which affect him both directly, by acting as human parasites, and indirectly, by damaging products of all sorts which he wishes to use for his own ends. The relationship of man to these pest animals is part of the more general topic of 'human ecology', and to appreciate more fully the present magnitude of pest control problems, the way they have arisen, and the future outlook we will look at this subject in a little more detail.

'Human ecology' is the term applied to 'the scientific study of the relationship of man to his environment', and in practical terms this means his relationship to his physical environment, his parasites, i.e., the phenomenon of human disease, and to his supplies (or resources) of every requirement such as land, water, minerals, energy, wildlife, food and so on.

It has been claimed that such ecological problems are the most serious dilemmas which man will have to face in the coming decades and their magnitude grows as man's demand for the natural resources of this planet expands. This demand for resources is proportional to the size of a particular human population, and the level of cultural development which it shows. Historically there have been three major stages in the development of human societies, first the food-collecting and hunting culture, next the farming, and finally the urbanised industrial culture. In popular language 'cultural level' can be largely equated with 'standard of living', and throughout the world, in developed and under-developed countries alike, there is an almost universal desire to increase standards of living to a maximum. The result is that whereas in the past man has generally produced an increasing number of cultures and sub-cultures during his evolutionary history, now, as Teilhard de Chardin* (1956) has suggested, this tendency has been reversed and there is a movement towards a broadly similar, world-wide, culture based on a scientific industrial civilisation. To illustrate what differences in 'cultural level' can mean in terms of health and consumption, Table 1 shows some contrasts between India, where in the main the people are poor, peasant agriculturalists, and the U.S.A., whose citizens have one of the highest standards of living in the world.

The second factor governing man's demand for resources is the human

* Teilhard de Chardin, P. 1956. The antiquity and world expansion of human culture. *Man's Role in Changing the Face of the Earth*. Ed. Thomas *et al.* Chicago, 103–112.

TABLE I SOME CONTRASTS BETWEEN TWO CULTURAL LEVELS

	India	*U.S.A.*
Mean expectation of life at birth in years	41·2 (1951–60)	70·0 (1963)*
Annual infant mortality rate of children under one year old per thousand live births	81·0 (1962)	25·3 (1962)*
Daily per capita food consumption in megajoules	8·1 (1962–63)	13·0 (1964)†
Annual per capita meat consumption in kilogrammes	1·0 (1962–63)	103·0 (1964)†
Annual per capita steel consumption in kilogrammes	16·0 (1964)	615·0 (1964)†
Annual per capita energy consumption in kilogrammes of coal equivalent	161·0 (1964)	8772·0 (1964)†

The date in brackets is the period for which the figures apply.
* United Nations Demographic Yearbook 1964.
† United Nations Statistical Yearbook 1965.

population size, and the history of the growth of the human population of the world, and its present explosive position is well known. Table 2 shows some figures illustrating this growth, together with the United Nations estimates of future trends to the year 2000. So far as we are concerned, the salient feature

TABLE 2 CHANGES IN THE HUMAN POPULATION OF THE WORLD (IN MILLIONS)

Date	*Population*	
1 A.D.	200	
1600	400	
1850	1100	
1900	1550	
1960	2998	
1980	4551	High Estimate
	4330	Medium Estimate
	4147	Low Estimate
2000	6994	High Estimate
	6130	Medium Estimate
	5449	Low Estimate

is that the present (1967) population of some 3000 million, which it has taken man the whole of his history to date to attain, can be expected to double by the end of the century.

It is now clear that we have a continual rise in the population, and in the cultural level in the developing countries as they industrialise. Similarly, in the developed countries the population is also rising, although at a slower rate, but here too there is a continual demand for further improvements in the standard of living, economic progress being largely equated with increasing consumption. Thus, as the world's population and the average cultural level both rise, we can confidently forecast an unprecedented and alarming rise in demand for the world's natural resources.

Pest control activities fit into this general picture in a complex way, as follows:

(*a*) A good deal of the present increase in world population can be attributed to the control of mortality factors. The average life expectation at birth in Britain in 1750 was 25–30 years, whilst it is now nearly treble this at just over 70 years. In this, pest control has played, and is still playing a large part. Thus, to cite one example, in the early nineteen fifties, 3 500 000 deaths a year throughout the world were directly caused by malaria, whereas in 1966 there were less than 1 000 000. Strenuous efforts are being made to eradicate the malarial parasite on a world-wide scale, chiefly by the control of its mosquito vectors.

(*b*) Increasing populations, and in particular the increased densities associated with city life, present particular medical problems, as disease transmission becomes much easier when people are in regular close contact. In many countries, bulk food storage and sewage disposal are primitive, and encourage the build up of large populations of commensal animals such as cockroaches, flies, some species of birds, rats and certain other mammals, producing further health hazards. Thus, in 1841, whilst the average life span in England and Wales was 41 years, and 45 years in the Surrey countryside, the mean expectation of life of people living in the London metropolis was 37 years, and in Liverpool 26 years.

(*c*) As standards of living increase, animal depredations become less readily tolerated by man. Fleas, lice and internal helminth parasites are a universal and normal part of everyday existence in many underdeveloped countries, but a comparative rarity, regarded with horror, in the highly developed parts of the world. In America, considerable sums of money are spent on the insecticide treatment of lakesides so that individuals fishing for recreation may be freed from the annoyance of harmless biting flies, whereas large parts of the underdeveloped areas of the world have yet to institute control procedures to limit the spread of mosquito-borne malaria. Malaria, as we have seen, is still responsible for almost a 1 000 000 deaths a year, and perhaps for a further 100 000 000 clinical cases of the disease. In order to achieve satisfactory health standards throughout the world, enormously increased efforts in pest control are still needed. It should be remembered that insect vectors are thought to be responsible for about one half of all human deaths and deformities due to disease. For example, at least 7 000 000 people in Latin America suffer from Chagas' disease, caused by an insect transmitted protozoan, *Schizotrypanum* [=*Trypanosoma*] *cruzi*, whilst in parts of Africa, onchocerciasis (an insect-borne nematode disease caused by *Onchocerca volvulus*) infects over 50% of the population, of which 30% have impaired vision and 4–10% are blind.

An exactly similar situation is found when we turn to foodstuffs and other materials which suffer from pest attack. In the highly developed countries a cockroach wingcase in a packet of rice or a loaf of bread is frequently the cause of legal action against the retailer; in many underdeveloped areas, on the other hand, it is customary to sieve the live weevils out of grain immediately prior to using it. Indeed where housewives are highly selective in their tastes, a good deal of pest control work centres around the need to market attractive-looking produce—apples without a blemish, potatoes without a single wireworm hole and lettuce without an aphid.

(*d*) Finally pest control has a most important part to play in satisfying the unprecedented increase in demand for resources of all sorts (agricultural crops, meat, fish, wood and textiles), which we have already noted. We hear increasingly frequent warnings of present food shortages and famines to come, and in his report on the state of food and agriculture in 1966, the Director General of the Food and Agriculture Organisation of the United Nations wrote: "Thus the world food situation is now more precarious than at any time since the period of acute shortage immediately after the Second World War". The data upon which this melancholy conclusion is based appear in Table 3, whilst in Table 4 target figures showing the staggering effort needed to have an adequate diet for all by the end of the century are presented.

TABLE 3 RECENT CHANGES IN THE PER CAPITA INDEX OF FOOD PRODUCTION
(In this Index, the average 1952/53 to 1956/57 is considered to be 100.)

Region	*Pre-World War II*	*55/56*	*60/61*	*65/66*
West Europe	93	100	112	118
North America	85	99	100	104
Latin America	105	99	101	99
Far East	107	102	108	104
Near East	93	98	105	101
Africa	100	98	103	96
World	95	100	107	107

(Food and Agricultural Organisation 1966)

TABLE 4 ESTIMATED PERCENTAGE INCREASES REQUIRED TO PROVIDE AN ADEQUATE DIET FOR EVERYONE

	Low calorie countries		*World as a whole*	
Year	*1975*	*2000*	*1975*	*2000*
Cereals	40	130	35	110
Pulses	100	275	85	225
Animal production	120	485	60	210
All foods	79	293	51	174

(Third world food survey 1963)

There is no single solution to this enormous problem. Ultimately completely new sources of food may be found, but the immediate problem can only be solved if new land is brought under the plough, and the productivity of existing land is increased by more efficient agricultural techniques, including the use of improved plant varieties, more fertilisers and modern methods of animal husbandry. The relevance of pest control to this crisis is obvious; there is simply no surplus for animal competitors to take, and yet it is extremely difficult to make any estimates of the overall losses to food production caused by pest attack. There are two main reasons for this. Firstly in many parts of the world there is no systematic recording of pest outbreaks and the damage caused, and secondly except in the most obvious situations we are usually unaware of the losses sustained as we have no knowledge of the potential yields of animal and plant products in the absence of pest damage. Estimates of pest losses have, however, been attempted at various times and some of these are quoted below to give an idea of the orders of magnitude involved.

Strickland estimated in 1965 that, without control measures, crop pests such as slugs, insects and rabbits in Britain would have ruined the equivalent of 7% of our farming land. Similar figures have been quoted for the U.S.A. and Canada. In the underdeveloped areas of the world the situation is often worse, and it has been claimed that complete locust control alone could increase world food production by 15%. World losses to grain in store are about 10%, of which a good deal is due to insect damage but this figure increases to between 25 and 50% in Central America. Table 5 gives recent estimates for world crop losses from insect attack and, as we can see, an overall figure

TABLE 5 WORLD ANNUAL CROP LOSSES FROM INSECT DAMAGE IN MILLION TONNES*

Commodity	*Actual production*	*Estimated loss*	*Loss as % of present production*
Wheat	265·5	17·8	7
Rice	232·0	120·7	55
Maize	218·5	44·0	20
Other cereals	245·1	21·2	9
Potatoes	270·8	23·8	9
Sugar beets and cane	694·6	228·4	33
Vegetables	201·7	23·4	12
Fruit	141·7	11·3	8
Stimulants	10·2	1·9	19
Oil crops	94·7	14·5	15
Fibre crops	16·0	3·0	19
Natural rubber	2·3	0·1	4
Total	2393·1	510·1	21%

* Cramer, H. H. (1967) Plant protection and world crop production. *Pflanz. Nach.* **20** (1).

of 21% is suggested. Losses in livestock production are just as serious. It has been suggested that, even in advanced countries total production might be increased by 15% if internal parasites were controlled. African horse sickness, an insect-borne virus which broke out in the Near East and South-west Asia, killed some 300 000 horses, and the effect of this on farmers losing their only draught animals is difficult to overestimate. In the four million square miles of Africa south of the Sahara there are many areas with few or no domesticated animals, as this comprises the tsetse-fly belt where animal trypanosomiasis is endemic. Finally, animal and human health may be intimately connected; for instance in many parts of Asia and Africa between 25 and 75% of older cattle sent for slaughter are infected with beef tapeworm (*Taenia saginata*), and since man is the definitive host, this situation is reflected in the human population which in turn may be as much as 90% parasitised.

It is perhaps easier to justify the study of pest control than it is to justify another book on the subject. The reader must realise that a wide range of human activities—agriculture, fisheries, forestry, horticulture, livestock husbandry, veterinary and human medicine and public health, textile manufacture and building, are all faced with pest problems. In most cases excellent textbooks are available which deal with control practice in a particular field, and some of these appear in the list for further reading at the end of this book. However, there is much in common in all these disciplines and the object of the present book is to attempt to identify some of the basic principles, techniques and problems underlying the control of pest animals.

Chapter One
PESTS AND THE MEANING OF PEST CONTROL

1.1 The Pest Concept

Most people have a general idea of what they mean by a 'pest', but to attempt a rigorous definition of the word is more difficult. The word has been derived via the French word 'peste' from the Latin 'pestis', meaning a plague, pestilence or contagious disease, and consequently its early use was largely as a description of any deadly epidemic disease, particularly bubonic plague—hence the phrase 'a pox', or 'a pest on you'. Subsequently, the meaning became extended to any thing or person that is noxious, destructive or troublesome, being used initially to describe troublesome people. Accordingly, it is clear that when an animal is described as a pest, a word of very general application is being used which cannot really be confined to some specialist scientific definition such as "Insects are pests when they are sufficiently numerous to cause economic losses" (DeBach, 1964).

In effect then, a pest animal is any animal which is noxious, destructive or troublesome to man, or to his interests.

Defined in this broad way, the list of animals one might regard as pests depends entirely on the point of view of the person concerned. To the farmer and horticulturist, the rabbit is a serious pest, and the flea which acts as a vector of myxomatosis a 'beneficial insect'. To the dealer in rabbit meat, and to the pet lover, the rabbit is the desirable animal, and the flea is the pest. The definition of the term 'weed' presents exactly similar problems and on this topic Stern (1956)* has written "Taken as a whole, weeds are not so much a botanical as a human psychological category within the plant kingdom, for a weed is simply a plant which in a particular place at a particular time arouses human dislike. . . ." A similar argument applies to pest animals.

In practice, however, pest animals are so important in some fields of human activity that special technologies have developed with their own staff and literature. Table 6 is an attempt to list the more important ones, and broadly defines the range of this study. In this book, 'Pest Animals' will be interpreted as including the relevant members of the Protozoa, and Metazoa, causing any form of damage to man or to the animals, plants or materials valued by him.

* Stearn, W. T. (1956) *J. Roy. Hort. Soc.*, **81**, 285–6.

TABLE 6 THE MORE IMPORTANT HUMAN ACTIVITIES AFFECTED BY PEST PROBLEMS

Activity	*Types of pests involved*	*Examples*	*Agents of large-scale control technology*
Crop husbandry, Horticulture and Forestry	Consumers, parasites and transmitters of crop diseases	Eelworms, snails, aphids, beetles, locusts and rabbits	Agricultural chemicals industries; Government crop protection divisions
Animal husbandry	Consumers, parasites and transmitters of livestock diseases	Protozoa, roundworms, tapeworms, fleas and foxes	Agricultural chemicals and drug industries; Government veterinary divisions
Fisheries	Consumers, parasites and transmitters of fish and shellfish diseases	Protozoa, flukes, starfish, roundworms, tapeworms	Government fisheries departments
Food storage and processing	Consumers and contaminants of stored products	Mites, cockroaches, weevils and rats	Disinfestation companies, and public health departments
Medicine	Human parasites	Protozoa, roundworms, tapeworms and lice	Medical services and the drug industry
Public health	Human commensals and disease transmitters	Mosquitoes, lice, bedbugs, rats and bats	Government public health division
Structure maintenance	Fouling agents, and consumers of wood, textiles and hides	Pigeons, barnacles, beetles, moths and rats	Disinfestation companies on behalf of architects, storers, builders, furniture makers, shipbuilders, and museum curators

One reservation which should be made to the definition of a pest animal as 'any animal which is noxious, destructive or troublesome to man, or to his interests' is that the target species for pest control, and hence by inference the 'pest', is not necessarily *directly* destructive or troublesome, although the term 'noxious' may be applicable. Thus a variety of animals such as wild game slaughtered in tsetse-fly control programmes in Africa, snails killed in the control of human schistosomiasis and sheep liver fluke (see also p. 36), and coccid-tending ants killed in the control of swollen shoot disease of cocoa (see also p. 54), are selected as 'target species' because of their linking role in a destructive chain of events which is most easily broken at this point.

1.2 Criteria for Assessing Pest Status

As defined above, the number of pest animal species is legion and to some extent a matter of personal taste. A more objective classification could be obtained by listing all the pests of a particular plant or animal crop in a particular area, and arranging them in the order in which their activities lower the final financial return from the area. It would then be possible by deciding on arbitrary levels of percentage loss to classify the pests into such categories as 'major economic importance', 'minor economic importance', and 'negligible economic importance'.

When it comes, however, to taking steps to counteract the noxious, destructive or troublesome habits of pests, a new criterion comes into play, viz: whether the damage and inconvenience is sufficient to warrant the trouble and expense of counter-measures. Under these circumstances we can classify some of our pests under the new heading of 'economically controllable pests'. In practice, the criteria for assessing 'economically controllable pest' status are only a little less subjective than those for assessing 'pest' status.

In matters of medicine and public health, we customarily think of human life as invaluable. By this criterion, medical and public health pests would automatically be economically controllable pests (assuming effective control measures were known). In practice this view becomes modified, depending upon who is paying; the man dying from malaria or Chagas' disease will spend his last penny in seeking a cure; large sums can be extracted from tax payers to support public health measures in their area, particularly after epidemics have broken out. On the other hand, relatively meagre sums are made available by the developed countries for work in underdeveloped ones, even when the likely saving in human lives and deformity as a result of campaigns against such diseases as malaria, sleeping sickness, onchocerciasis and Chagas' disease can be estimated beforehand with some accuracy. In general, however, it is true to say that pests of medical importance are given high economic priority. Man is prepared to tolerate very little physical or psychological damage from them, and as a consequence, complete eradication of the pest is a common aim.

Most other pest damage is damage to property rather than person, and here a more objective valuation of the loss is possible except perhaps in such cases as pets threatened with disease, or works of artistic or sentimental value threatened with beetle and moth.

At first sight, the decision regarding the economic status of a pest attack on a plant or animal crop would seem a fairly simple matter, the cost of the expected pest damage would be assessed, as would the cost of counter-measures to prevent it. If the damage cost was greater, then the pest would be economically worthwhile controlling and the necessary measures would be instituted at once. In other words we must ask, "Is the expression

$$\frac{\text{Cost of preventable pest damage}}{\text{Cost of control measures}}$$

(the cost:potential benefit ratio of Ordish 1952) appreciably greater than unity?"

Closer examination shows the situation to be much more complex, rendering cost:potential benefit ratios calculable in only a few cases, and even then on a short term view only. Some of the factors involved in assessing a cost: potential benefit ratio are as follows.

(i) If control measures are undertaken, it is to prevent further damage, whether or not actual damage has already been caused. In most cases, the estimation of future damage is a hazardous procedure as pest attacks frequently fluctuate unexpectedly for reasons often unsuspected at the time. Thus, it has been estimated that 75% of the fields treated for wireworms would probably not have produced damaged potatoes even if left untreated.

(ii) In the case of long term 'crops' such as forests, livestock kept for breeding, and the long-term storage of materials, estimates of the likely recurrence of pest attack are required. Thus, a single treatment might be economically justifiable; if in the end, however, 3 treatments have been required before 'harvest', the result might be a substantial loss.

(iii) The likely effects of any control measures on the pest cannot always be predicted with any great precision, owing to such variables as pest resistance and the effects of environmental factors, such as climate and the physical situation, which prevent the control measures from being fully effective.

(iv) The costs of specified control measures can usually be assessed with some accuracy, but of course can sometimes be upset by changes in the costs of materials and labour, or by encountering unexpected difficulties or delays during control operations.

(v) Labour, resources and skill may not be available when control measures are required, or in some cases may take so long to assemble that irreparable damage has already resulted, a problem particularly serious in the prevention of locust swarms (see Plate 1A) from forming by the destruction of hoppers in remote areas.

(vi) Not infrequently, the control measures themselves cause damage to the crop or animal through toxicity, and by mechanical damage to crops when bulky spray equipment is used in fields. In extreme cases, the cure becomes more damaging than the disease.

(vii) Control measures may have many side effects, some advantageous, as when pests other than the target species are controlled, or when they actually stimulate growth, or otherwise improve the selling features of a product, but others are disadvantageous. Control addiction can be found wherever pests are troublesome, and results either from a 'security complex', by which an initially successful treatment is needlessly repeated time after time 'just to be on the safe side'—psychological addiction; or from the damage which the control procedure inflicts on the parasites, predators and competitors of the pest species which thereby makes pest control measures more necessary than before—ecological addiction. Other side effects such as the creation of new pests and the effect of treatment on other non-pest species, particularly wild life, all carry economic implications.

(viii) Finally, except under conditions of guaranteed prices which merely shift the burden from the individual to the community, the value of the commodity being protected is liable to wide fluctuations. Indeed, in a market adequately supplied with a particular product, a sudden saving of 10% of the crop by pest control measures would simply lead in the short run to over production, and a consequent drop in price. Thus, estimates made in 1938 for various common crops, showed that an increase in production of 1% produced a price depression which ranged from 1·4% for cotton to 3·3% for potatoes. The economics of this situation are complex, and are particularly influenced by the elasticity in demand for a particular crop. In the United Kingdom, strawberries are an example of a crop showing considerable demand elasticity, where a small price decrease stimulates a considerable increase in consumption. Potatoes on the other hand, are relatively inelastic and small price decreases do not usually stimulate the consumer into eating more generous portions. It is with these inelastic demand crops that there is a real danger of effective pest control measures reducing the monetary returns to the farmer in the short term.

As the cost:potential benefit ratio is based on so many imponderables, it is not surprising that at present, decisions on 'economically controllable pest status' are more of an art, relying on experience, than a science relying on calculation, when the economically successful pest control officer has to possess many of the attributes of a successful gambler. As cost:potential benefit ratios are extremely variable, depending upon the pest and the control measure adopted, there may or may not be latitude for error, and the long-term objective must remain the replacement of imponderables by accurate parameters, so giving the assessment of pest status as objective a basis as possible.

1.3 Pest Origins

On the assumption that prevention is better than cure, interest has been shown in recent years in how economic pest species arise. From our definition of a pest, it is of course a truism to say that all pests are created when man recognises them to be such. Thus, the malarial parasites, *Plasmodium* spp. became recognised as pests on November 6 1880, when Laveran discovered that malaria, perhaps then the most widespread of all human diseases, was caused by these protozoan parasites. Eighteen years later, Ross and Grossi independently demonstrated the importance of mosquitoes in the life cycle of malarial parasites, so elevating the anopheline mosquito to the status of a pest of overriding medical importance. As knowledge increases, pest lists grow longer. Pest origins can be classified as follows.

1.3.1 Naturally occurring pests

Most human parasites seem to be of long standing, and doubtless many may have evolved in association with our 'pre-*sapiens*' ancestors. This must similarly be true of many pests of livestock, as there are reports in the literature of epidemics occurring among free-living wild animals and birds. Again, certain plants growing in natural or semi-natural environments suffer periodic plagues of destructive insects; virgin stands of the balsam fir (*Abies balsamea*) in Canada are periodically decimated by a small moth caterpillar, the spruce budworm (*Choristoneura fumiferana*) (see Fig. 3), and a study of the annual rings of surviving trees suggests that this outbreak pattern has persisted for as long as we can detect. Locust outbreaks, though perhaps encouraged by certain human land-use practices, are probably a normal feature of desert regions in certain parts of the world as are eruptions of lemmings (small rodents) in the Arctic.

In short, there are pests which seem to have plagued mankind from the beginning, and are part of the natural biological complex. From our knowledge of evolutionary processes, we can expect that other pests will arise from time to time. A probable example of this in Britain is an apple capsid bug *Plesiocoris rugicollis* which prior to 1900 fed on willows, and was not recorded from apples. However, not all pests are of this sort; perhaps the majority are 'man-made'.

1.3.2 Pests arising from man's increasing demands

As standards of living increase, so expectations are heightened for human health and comfort, for crop yields from a given area, for meat and milk yields from a given animal, and also for the type and quality of agricultural produce.

The tarnished plant bug (*Lygus pratensis*) is an economic pest chiefly by virtue of the damage it causes to fruit, and in particular apples. This damage is almost entirely 'cosmetic', (blemishes affecting the appearance of the fruit) and as a consequence this pest has arisen largely from the visual preferences of housewives. Any increase in the value of a commodity is liable to increase the number of economically important and economically controllable pests, and increasing pressures for both food and land have this effect.

1.3.3 Pests arising from man's alterations to the ecosystem

One of the most characteristic features of man is his ability to manipulate his environment in the hope of suiting his own ends. Unfortunately, it seems that in the process he has increased the depredations of pest animals, a situation which has developed for a variety of reasons.

(a) CHANGES IN SOCIAL HABITS

Pests of medical importance are particularly affected by changes in man's social habits. Civilisation, with its dense aggregates of people often housed in cramped and insanitary conditions has produced its own pest problems, especially in the multiplication of commensal animals, and in facilitating disease transmission. The large-scale use of wood in housing and furniture, and of cotton, wool and leather in clothing and upholstery means that the animals which in their original field habitat were associated with the breakdown of these materials have become important pests. Any new human activity may bring man into conflict with a new range of pests, and any large scale exploitation of a new food source or raw material is likely to provide opportunities for similar exploitation by animals. The first insect to evolve a polythene feeding habit will strike a rich vein!

(b) PLANT AND ANIMAL BREEDING

Through selective breeding, man has endeavoured to improve both the quality and yield of the food and raw materials which he obtains from animals and plants. In selecting for better yielding strains, he may, however, alter the genetical resistance of species to pest attack, thereby creating new pests. This variable susceptibility may occur in cultivated plants as seen in the damage caused to varieties of wheat by the hessian fly (*Mayetiola destructor*), a stem-boring dipteran pest of particular importance in North America. The fly-resistant wheat variety Pawnee was developed after 14 years of research beginning in 1928, and was the result of a cross between the resistant variety Kawvale, and the susceptible one Tenmarq. Certain varieties of oats (Picton, Pennant and Milford) are more resistant to stem eelworm (*Ditylenchus dipsaci*) than are others (see pp. 45–47 for further discussion of eelworms), whilst certain breeds of domestic cattle appear to be more resistant to gut nematode parasites than others, a genetic line of Nigerian Zebu cattle being relatively resistant to several species. The fact that deliberate selective breeding can alter resistance levels shows that it is quite possible that cultivated species might be more susceptible than their wild ancestors.

(c) THE INTRODUCTION OF PLANTS AND ANIMALS

In 1876, Alfred Russel Wallace pointed out in his book *The Geographical Distribution of Animals* that major discontinuities existed in the world's flora and fauna. That is to say that many species of plants and animals through isolation have been denied the opportunity of spreading to seemingly favourable parts of the world. In the last hundred years or so, man's increasing mobility has provided that opportunity, and Elton (1958) in *The Ecology of Invasions by Animals and Plants* has given a stimulating account of the process. This 'historical convulsion in the world's fauna and flora' as he calls it has provided perhaps the greatest source of new pests. This can occur (i) by the introduction of an animal into a new area, where it either feeds on local useful plant or animal hosts and so achieves pest status; or, (ii) by the introduction of a useful animal or plant, which provides previously innocuous members of the local fauna with a new food source and so turns them into pests.

In the case of parasitic diseases, those that are endemic may cause little observable damage to the local host animals, being maintained at a subclinical level by immune mechanisms. However, the introduction of susceptible animals (animals that have not previously experienced infection with those parasites) from a new area may completely alter the economic status of the parasite.

To anticipate a later section of this chapter, there is frequently a connection between economically controllable pest status and the numbers of pest individuals, and two factors in introductions are especially likely to permit population increases. Firstly, the introduction may provide a new and abundant food supply; a new 'living space', as Moses found when he led the Israelites to the promised land, or when the pioneers first sighted the great American prairies. Secondly, during the process of introduction, predators and parasites which previously had regulated numbers, may have been left behind.

Whatever the reason, many pest animals originate in this way. Thus, it has been estimated that in the U.S.A. about one half of the major animal pests of plants have been imported, whilst the number of imported plant species and varieties on which local pests could arise has been put at some 200 000. From the wealth of examples of pests arising through plant and animal introduction, only three examples will be quoted here. Firstly, maize (*Zea mays*) almost certainly originated in Central America, although it is now the most widely distributed cereal in the world. An examination of some of the more important insect pests (see Table 7) of this crop shows that during its spread local insects in the new localities have assumed the status of pests. Although 'ecological equivalents', the species are not necessarily closely related.

Secondly, the colorado beetle (*Leptinotarsa decemlineata*) when first dis-

TABLE 7 SOME INSECT PESTS OF MAIZE; A REGIONAL COMPARISON

Pest type	*Species*		
	South America	*Africa*	*India*
Crickets destroying young plants	*Scapteriscus vicinus*	*Gryllotalpa africana*	*Gryllulus domesticus*
Caterpillars attacking foliage	*Laphygma frugiperda*	*Cirphis unipuncta*	*Spodoptera mauritia*
Stalk boring caterpillars	*Diatraea* spp.	*Busseola fusca*	*Chilo zonellus*
Cob boring caterpillars	*Heliothis zea*	*Heliothis armigera*	*Heliothis armigera*
Disease-transmitting Hemiptera	*Baldulus maidis*	*Cicadulina mbila*	*Cicadulina mbila*

covered in 1823 was an inconspicuous beetle feeding on the buffalo bur (*Solanum rostratum*), and remained so until the early pioneers introduced the potato (*Solanum tuberosum*) into the area in about 1859. This provided a new and suitable source of food, and the beetle began to spread eastwards through the potato-growing areas reaching the Atlantic in 1874. Several accidental introductions occurred in Europe, but all were eradicated until 1920, when control measures failed at Bordeaux (see Fig. 1). Large areas of Europe are now infested, and the cost of control measures is reflected in increased potato prices.

Thirdly, a blood-fluke of man (*Schistosoma mansoni*) which is found in Equatorial Africa, Malagasi, Egypt and S.W. Arabia, was transported, probably by African slaves to South and Central America. There it managed to find a suitable snail intermediate host and has now spread through a large part of Brazil, Venezuela, Guyana and the West Indies. Another parasite, the hookworm of the new world (*Necator americana*) probably reached North America in the same way. Indeed, there are numerous examples of animal parasites being transported by carrier animals, in animal produce, or within vectors from one part of the world to another, there to establish infections.

Although a great many pests have originated by introductions, there are probably many more awaiting the opportunity. For example, leaf-cutting ants of the genera *Atta* and *Acromyrmex*, sometimes considered to be the most serious general insect pests of the New World Tropics, are not known in the Old World, but since they use a wide range of leaves simply as a compost for fungus culture, it seems possible that, given the opportunity, they could become just as serious pests in Africa and Asia.

(d) THE PRACTICE OF MONOCULTURE

Hunting wild animals in the forests for meat, and searching out roots, seeds and berries from among a mass of vegetation are inefficient methods of getting food, because the supply is erratic, and a good deal of energy is required for collection as useful food items are few and far between. This was changed with the advent of agriculture some 10 000 years ago, as man cut down the forest, and began increasingly to plant large areas with one type of useful plant, or to keep large dense herds of one type of animal, restricted in their movements. A large aggregation of one species of plant or of one species of animal is called a 'monoculture'. It is possible to have either (a) a single species culture in *space*, i.e., uniform over a wide area, or (b) a monoculture in *time*, where a given species is reared in the same place year after year, regardless of the number of individuals involved. In this section, we are mainly concerned with spatial monocultures, but temporal monocultures also produce serious pest problems. (See Section 2.2.5.) Examples of stages in increasing spatial monoculture in agriculture are (i) The traditional 'garden' of the tropical peasant farmer in its patch of newly cleared bush, where individual plants of coffee, cocoa, cassava, tannia, plantain, pigeon peas, peppers, and often maize are scattered about haphazardly, with one or two animals

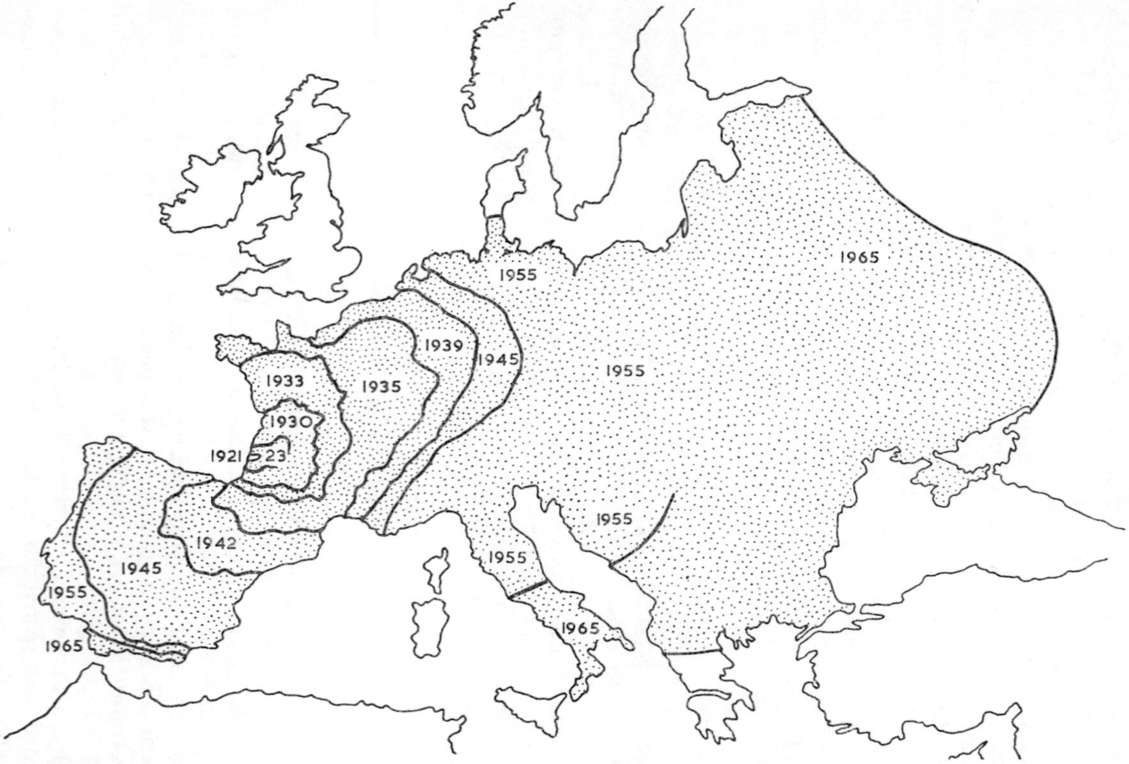

Fig. 1 Approximate lines of advance of the Colorado Beetle in Europe since 1921. N.B. Records for infested Eastern European countries were not available for 1955.

staked out for good measure. (ii) The temperate market garden with beds of only a few hundred square metres. (iii) The enormous farms and ranches of North America and Australia, where single stands of crops often exceed 80 ha. Figures for average field sizes are difficult to obtain, but an index of this can be found in the average farm size, which in 1965 was quoted as being Indonesia, 1 ha; India, 2·5–3 ha; Italy, 6·2 ha; U.K., 42·5 ha and the U.S.A., 122·5 ha. The economics of increasing mechanisation favour further increases in crop stand and herd sizes (a greater degree of monoculture), so that in 1920, the U.S.A. had 67 000 farms larger than 400 ha, whilst the figure for 1954 was 131 000.

Large scale monoculture with its concomitant large scale ploughing, planting and harvesting involves the replacement of complex assemblages of animals and plants (complex communities), by simple ones. Thus, an area of complex tropical rain forest, with perhaps 100 different types of trees in 2 acres, is replaced by an almost pure stand of sugar cane, and the 'plains game' of Africa with its great variety of species, is replaced by domestic cattle.

This reduction in community complexity is illustrated by a study of the effects of agriculture in the steppes of the U.S.S.R. where Bey-Bienko* (1961) found 330 species of insects in virgin steppe, and only 142 in wheat fields. The number of individual insects, on the other hand, showed the opposite tendency; 199 per m^2 in virgin steppe, and 351 per m^2 in wheat. In other words, monoculture produced a decline in the number of animal species, but an increase in the number of individuals of those species able to utilise the crop plants; these were mainly important pests.

The main effects of monoculture on potential pests are:

(i) a reduction in the number of predator, parasite and competitor species. This point is particularly emphasised in the Russian work, where the reduction was greatest, as a result of direct destruction, in the newly cultivated fields. Predator numbers may also be affected because of increasing dependence on the numbers of a particular prey. A wide range of prey gives the predator increased flexibility so that feeding habits can be altered if one food source becomes scarce. Prior to 1954, rabbits comprised some 55% of the fox's diet in the U.K. After myxomatosis, however, and the virtual eradication of the rabbit, the fox was able to exploit more fully alternative food sources, such as field voles and brown rats. Monoculture tends to restrict the range of prey available to a predator.

(ii) easier host location. In many animals a very large number of individuals die during dispersion as is illustrated for the spruce budworm (*Choristoneura fumiferana*) in Table 8, where total mortality from this cause was 67% and 75% in the two areas quoted. There may be several reasons for these high mortalities such as increased susceptibility to predators, and to climatic factors, but the majority probably die because they fail to find a suitable new

* Bey-Bienko, G. YA (1961) Quoted in Uvarov, B. P. (1967) Pest Articles and News Summaries **13**, (3), 202–213.

host. This is borne out by reports of budworm larvae being found dead on the surface of lakes in densities of 1000 per square metre, and of adult moths settling out in town centres in such numbers as to resemble snowstorms. The mortality figures quoted are for relatively pure stands of balsam fir, and in mixed forests it is conceivable that mortality would have been even higher. In situations where hosts are well scattered, and interspersed with non-host organisms, many potential pests such as insects and helminths must die without ever locating a host. When animal parasites of plants fail to find a host at the end of the dispersal phase, they die, as do endo-parasites of animals when ingested by a non-host. Monoculture must dramatically reduce dispersion losses, and also enhance the chances of locating both a mate, and suitable egg-laying sites.

(iii) a decrease of potential food stocks for the general feeders. When a large scale monoculture replaces a natural climax community, there is commonly a considerable reduction in the 'stock' or 'standing crop' of animals or plants, i.e., the biomass at any one time is reduced. Long-living general feeders may then find themselves short of food material, and so concentrate with disastrous effect on the cultivated crop. The large wireworm populations characteristic of grass leys gradually decrease with cultivation due to a variety of factors, but in the first year or so they may decimate cereal crops. Again there is evidence to suggest that the leaf-cutting ant *Atta cephalotes* when in its natural habitat of tropical rain forest, has developed a grazing mechanism which evens out its activity over the area surrounding its nest, so preventing over-exploitation. When citrus is planted after the forest has been cut down, the much smaller volume of foliage may be devastated by these long-lived colonies.

(iv) alterations to the microclimate. Because of the general reduction in the standing crop of vegetation noted above, the microclimate is less insulated from the greater fluctuations in the macroclimate than it was in a climax community. Consequently, the microclimate now fluctuates more, and these fluctuations are intensified by the periodic harvesting of the crop.

All these effects of monoculture, reinforced by frequent changes of practice, crops, and stock lead to instability, in the sense that the numbers of a given animal species, especially a potential pest species, are liable to change more dramatically from season to season, and from year to year than they would in the more complex climax community which monoculture replaced. Thus in such a situation pests are more likely to arise.

In fact, this theory (the stability of the heterogeneous state) is more generally applicable, and states that complex communities are more stable and less likely to have pest problems than simple ones, and there is field evidence to bear this out. Our examples of animals liable to disruptive outbreaks come from simple, inhospitable communities (spruce budworm in pure forest stands, lemmings in the arctic, locusts from the deserts); not from complex communities such as tropical rain forests, where pest outbreaks are virtually unknown. The present road of 'simplification for efficiency' may be a very dangerous one.

(e) THE USE OF TOXIC CHEMICALS

The use of toxic chemicals in pest control is perhaps worthy of separate mention, as it is a measure deliberately designed to kill a portion of the fauna, thus simplifying the community. As this is a special case of the more general phenomenon just discussed, we might expect new pests to arise as a result, and this has indeed happened, the most famous example being the sudden rise of the fruit-tree red spider mite (*Panonychus* [=*Metatetranychus*] *ulmi*) to pest status in the late 1940s due to the intensive spraying of orchard trees with DDT and other contact insecticides. These seem to have had a much more profound effect on the 45 or so species of predators than on the mite itself, hence the population upsurge.

It will be clear that there are many possible reasons why pests arise. To some extent prevention is possible, as in the case of quarantine measures to prevent the spread of undesirable species. In other cases, especially in the increasing demand for better quality and greater quantity of food, and in the continuing trend towards monoculture, and reliance on chemical pesticides, the difficulties of preventing pest outbreaks arising are going to become more serious. The changed conditions which may permit one species to assume the status of a pest may concomitantly bring about the demise of another. Occasionally, the declining species may be a pest but unfortunately it will more often be a harmless, or perhaps a useful animal.

1.4 The Control of Damage

Already, the word 'control' has been used a good deal in this book, and it has been used in a variety of senses. Here we are concerned with its use to cover the idea of 'regulating'. In practice, people suffering from the depredations of pest animals are interested in regulating the extent of damage they cause both now and in the future. Indeed, wherever possible they would like immediate damage to cease altogether, and future devastation to be prevented, with the result that in the pest industry, the concepts of regulation (or control) and eradication have become confused.

As the reduction, and if possible prevention of damage to hosts is our principal aim, it is worth looking in a little more detail at the factors which determine the extent of pest damage. For a given pest species, these are broadly as follows:

(a) The destructive power of the individual pest. Amongst other things, this may vary with the pest's age and size (the first instar larva of a clothes moth does less damage than the final instar larva, whilst the adult which cannot digest wool does no damage); with the pest's genetic constitution (the damage done to red clover by the stem and bulb eelworm *Ditylenchus dipsaci* depends upon which of the five potentially destructive biological races is involved); and with the pest's environment (the damage done by rabbits to the bark of young trees largely depends on the non-availability of other herbage, especially in winter).

(b) The length of time that the host is exposed to pest attack. This can be varied either by altering the duration of the host's susceptible period, or by reducing the overall time that the host is exposed to damage. An example of the former is provided by flea beetles (*Phyllotreta* spp.) which are most injurious to cruciferous plants during the seedling stage, and the more quickly this stage can be passed through by addition of fertilisers and improved tillage, the less will be the resultant damage. Again, the longer sheep are infected with the liver fluke, *Fasciola hepatica* (see p. 36) before treatment, the greater will be the resultant damage.

(c) The resistance of the host. Different individuals of the same host species vary considerably in their susceptibility to pest attack, and hence in the degree of damage they exhibit. Obviously many factors are involved, but they include the individual's genetic constitution, its physical environment, its health, nutritional status, age, and previous history of attack (see also section 2.2).

(d) The number (population) of pest animals involved. Thus a wheat-field where the numbers of wireworms never exceeded 750 000 larvae per ha, or a flock of fowls with less than a hundred poultry mites (*Dermanyssus gallinae*) per bird, do not suffer significant damage.

As host damage is related to the foregoing four factors, it follows that, in theory at least, host damage can be reduced by manipulating any of them. In practice, however, it is perhaps fair to say that in the past, most attention has been paid to attempts to reduce the populations of pest animals. Now it is clear that there will not usually be any simple relationship between direct damage and pest numbers, and even if we were to conceive of a situation in which factors a, b, and c above were held constant, the relationship would rarely be linear (10 times the number of pests producing 10 times the amount of damage), in some cases it may be logarithmic, the amount of damage varying with the logarithm of the pest numbers. Thus, as can be seen in Fig. 2, the points relating the initial population density of larvae of the potato root eelworm (*Heterodera rostochiensis*) to potato growth at 38 days, lie on a curve which is approximately logarithmic (exponential) in form. Thus when we plot the logarithm of larval numbers against potato growth, an approximately straight-line or linear relationship results.

This is not surprising, for the pest is exploiting a resource (usually a food resource), and this sort of exploitation commonly obeys the law of diminishing returns, itself logarithmic in form. Thus the first insects on a crop can occupy the best positions and feed on the choicest material; subsequent invaders on the whole get poorer positions, poorer food supplies, and do less damage.

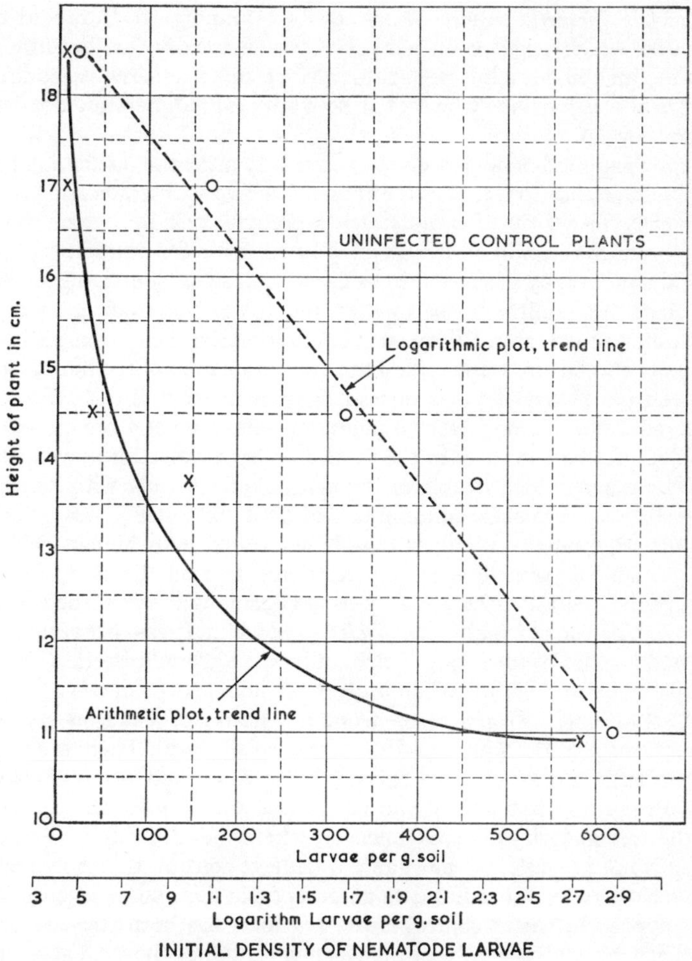

Fig. 2 The relationship between the initial population level of the nematode *Heterodera rostochiensis* and the subsequent growth of potatoes as assessed at the 38th day.

Even this is an oversimplification however, as many animals and plants can tolerate small numbers of some pests without manifest disease symptoms, maintaining them out of surplus production; a situation characteristic of well-adapted host-parasite relationships. Only when numbers exceed a threshold where compensation is no longer possible does damage become apparent, although at very low densities of some eelworms, increases in crop yields have even been recorded (Fig. 2) if the assumption is made here that height is synonymous with yield. In certain subspecies of the helminth parasite of sheep

Haemonchus contortus, there seems to be a numerical threshold during reinfection which triggers off a hypersensitive reaction in the host, which expels the incumbent adult population. Much more research is required into the types of relationships which exist between pest numbers and the degree of damage they cause.

Although applied biologists do not commonly make the distinction between control and eradication, it is a useful one. To deal with eradication first, this term means 'to get rid of completely' (literally to tear up by the roots) and although the eradication of damage may be achieved for example by breeding a resistant variety, by administering a curative drug or by erecting a good wire fence around a poultry house to keep out foxes, the eradication of a pest population is much rarer. Where such eradication has been accomplished there are usually special circumstances. Thus the elimination of the human malarial parasites from 11 out of 35 countries originally infested in the Americas, and of the mosquito vector (*Aedes aegypti*) of yellow fever from 21 out of 50 originally infested countries in the same area has involved heavy expenditure of public money which is not readily available for coping with non-medical pest problems. Moreover, continuous and close surveillance was possible because the yellow-fever vector is closely associated with human habitations, and the malarial parasite is largely restricted to man. Recently introduced pests may be eradicated before they have become well established, examples being the African mosquito *Anopheles gambiae* in Brazil between 1938 and 1940, and a series of invasions of Britain by the Colorado Beetle. The release of sterile males has eliminated an indigenous insect (the screw worm *Cochliomyia* [=*Callitroga*] *hominivorax*) from the island of Curaçao and from a substantial area of the southern U.S.A. (see section 6.2). There are many examples of large and conspicuous animals being eradicated, usually by hunting, and in Britain the last wolf (*Canis lupus*) was killed in Scotland in 1743.

In the vast majority of cases however, where the eradication of damage or of pests is not possible, we must aim at damage control, that is its regulation at a level where loss is reduced to an acceptable minimum. The importance of this approach, where control is the key idea, has been stressed by Geier (1966), and his concept of 'pest management' implies the "... acceptance of the continued existence of potentially harmful species, albeit at tolerable levels of abundance."

1.4.1 Population growth

All animals are inherently capable of producing more young than are required to replace them when they die, so that all animal populations would increase very rapidly if conditions for reproduction were good, and old age was the only cause of death. The potential power of increase of a population of animals is measured by the parameter 'r', this being the 'intrinsic rate of natural increase', or the 'innate capacity for increase'. Positive values of 'r' can produce very great population increases. Darwin estimated that one pair of elephants, notoriously slow breeders, could theoretically give rise to 19 000 000 offspring in 750 years; an individual oyster can produce up to 114 million eggs, the tapeworm *Taenia saginata* up to 10 proglottids a day each containing some 8000 eggs, and the common housefly *Musca domestica* can lay some 800 eggs in a lifetime. It becomes apparent from this that given sufficient time, any species could overwhelm the world with its numbers, although in practice, this does not happen. For a particular species 'r' varies with different circumstances, but nevertheless it is a useful parameter for comparing the potential powers of increase of different species, and of the same species in different environments. Predictions based on such information could be of considerable importance in pest control.

1.4.2 Control of population growth

Although well over 90% of all animal species that have ever lived on earth are now extinct, so that it would appear on a long-time span that extinction is the usual fate of a species, within the period of a human life, extinction is a rare event. If populations exist between the extremes of extinction and unbridled increase, then at this crude level, controlling factors must be operating. When we wish to study this subject in more detail, however, it becomes more controversial, and the first difficulty is that during a particular time interval, and in a particular place, a particular population of animals might not be under any control. It might be plummeting towards extinction, a situation with which conservationists are frequently concerned, or it might *apparently* be heading towards excessive numbers, a situation which is distressingly common to the pest control expert. Among the possible reasons for these uncontrolled situations are:

(i) The time factor. We may be watching a population in numerical transition because its habitat has been altered, or alternatively its particular control mechanism may involve a periodic disastrous reduction in numbers, followed by an equally dramatic rise, as has been described for the brown lemming (*Lemmus trimueronatus*) in Alaska, and our short period of observation may reveal only part of the cycle.

(ii) The population size factor. It may well be that the world population of, for example, the peach and potato aphid *Myzus persicae*, remains reasonably constant, but the population on a single potato plant is liable to the most wild fluctuations, including extinction if a predator should find it. In other words the further back we stand from the phenomenon and the longer we watch it, the more apparent does control become.

(iii) The habitat factor. A population at the very limit of the species' range, may be in a region where permanent habitation is not possible. Thus in good years for seed production in the rush *Juncus squarrosus*, the moth *Coleophora alticolella*, whose larvae feed on the seed capsules, extends its range up the mountain sides in Northern England. After one of the periodic severe winters, however, the rush does not set seed at the higher altitudes, and consequently the high altitude moth population, which may have been the one we were examining, is wiped out.

The finding of a population which fluctuates within narrow limits requires an explanation for its relative stability, and a useful analogy is the operation of a thermostat in a refrigerator. In this example, when a critical level of temperature is reached, a device switches on the motor. This reduces the temperature to a new pre-set level when the motor is switched off, and the temperature in the cabinet begins to rise again to its original level. By this means it is possible to regulate the refrigerator temperature within fine limits. The principle involved here is referred to as 'negative feed back', as information about the cabinet temperature, when processed by the thermostat is always used to reverse or 'negate' the previous trend.

Similarly animal populations have a built-in tendency to rise in numbers and ecologists have sought to explain the controlling factors for this tendency by negative feedback. In particular, they have looked for mortality factors which either begin to operate, or operate more severely when the population is high, and cease operation, or act less severely, when it is low. Such factors have been referred to as 'density dependent factors' and the critical test for their existence is that a reduction in the size of the population is followed by a more rapid rate of increase than occurred originally. Among the many factors responsible for mortality in natural animal populations are starvation, accidents, and the effects of predators, parasites, toxic substances, exposure to unsuitable weather, diseases, old age and overcrowding. Most of these factors also affect reproduction, and by lowering it, further restrain population growth. Whilst many of these factors act directly, causing mortality or reproductive failure which is imposed on the animals by outside agencies (extrinsic factors), this is by no means invariably the case, and Wynne-Edwards (1962) has stressed that through social behaviour, some species have evolved mechanisms which regulate their population growth from within by self-imposed mortality or reproductive failure. Such mortality factors originating within the animals' own behaviour patterns are termed intrinsic factors, and territory-holding in many species of insects, birds and mammals may be a mechanism of this sort, where surplus individuals are either killed, or are driven out (and so have nowhere to breed) by members of the animal's own species.

Generally, wherever populations are reasonably stable, then either extrinsic or intrinsic density dependent factors are involved, either causing mortality or depressing reproduction. The alternative is that randomly acting mortality factors unrelated to the population size happen in the long run to have killed on average precisely the average surplus of young produced. This is an unlikely coincidence which need not concern us. Unfortunately almost all mortality factors can in effect, if not directly, act in a density dependent manner so that a practical guide for sorting out the roles of the various mortality factors in the population in question is still lacking. Accordingly, the applied biologist, be he concerned with conservation or with pest control, cannot, without prolonged research, tell if the population concerning him is under natural control or not, nor which, if any, are the regulating factors. Confronted with the necessity of making managemental decisions he may then be forgiven if he unfairly thinks of general theories of animal population control as essentially academic, containing truisms and of little predictive value.

Insurance companies are very concerned with estimating the longevity of men and women for the profitability of their life assurance policies depends upon the precision of their calculations. They are also concerned with the relative importance of various mortality factors when they have to insure against specific risks, so that actuaries who are responsible for the calculations, have developed certain practical techniques for analysing the relative importance of different mortality factors at different ages. These techniques are now proving to be of considerable value to the applied biologist; in particular the use of 'life tables'. In Table 8, a life table is presented for a natural population of the spruce budworm (*Choristoneura fumiferana*). This insect is a serious pest of fir trees in North America, where the defoliating activities of the larvae kill many hundreds of acres of trees. As an analytical tool, the life table has many uses; it is simply a list of the mortality factors operating at each stage in the life cycle (see Fig. 3) and their relative importance as judged by the number of deaths each causes.

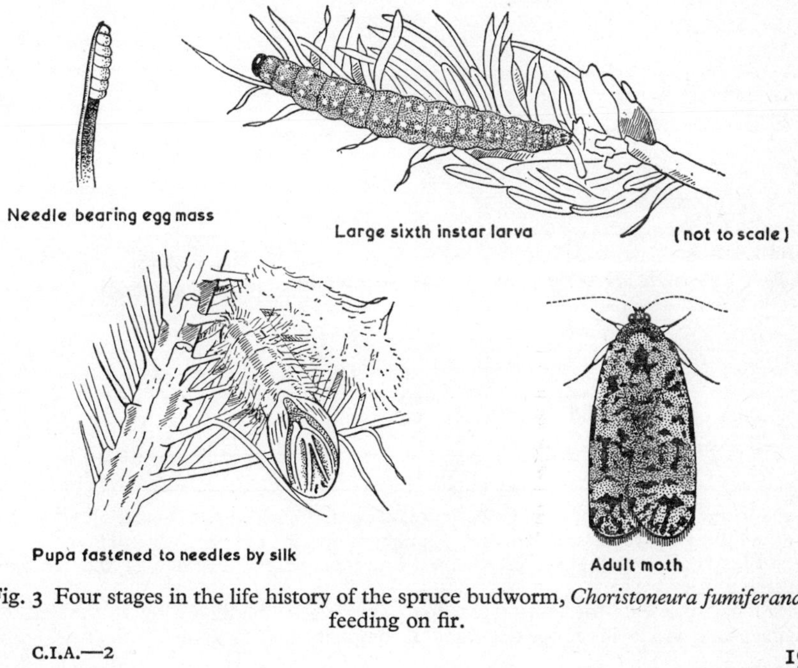

Fig. 3 Four stages in the life history of the spruce budworm, *Choristoneura fumiferana*, feeding on fir.

TABLE 8 LIFE TABLES FOR THE 1952–53 GENERATION OF THE SPRUCE BUDWORM *Choristoneura fumiferana* IN AN OUTBREAK, AND IN A NON-OUTBREAK AREA IN NEW BRUNSWICK

Modified from Morris and Miller 1954[1]

		NON OUTBREAK AREA		OUTBREAK AREA	
Age interval (X)	*Factor causing dX (dXF)*	*Number alive at beginning of X*(lX)*	*Number dying during X* (dX)*	*Number alive at beginning of X*(lX)*	*Number dying during X* (dX)*
Eggs		174		2176	
	Parasites		3		1
	Predators		15		174
	Unknown		1		21
	Total		19		196
Instar 1		155		1980	
	Dispersion, etc.		74·4		1148
Hibernaculae (Dormant over-wintering stage)		80·6		832	
	Winter		13·7		141
Instar 2		66·9		691	
	Dispersion, etc.		42·2		484
Instars 3–6		24·7		207	
	Parasites		8·92		2·9
	Disease		0·54		0·3
	Birds		3·39		1·7
	Starvation		—		165·3
	DDT		—		8·3
	Unknown		10·57		26·7
	Total		23·42		205·2
Pupae		1·28		1·80	
	Parasites		0·10		0·13
	Predators		0·13		0·11
	Unknown		0·23		0·27
	Total		0·46		0·51
Adult moths		0·82		1·29	
	Sex†		0		0·10
	Size‡		0		0·57
	Total		0		0·67
'Normal' adult moths		0·82		0·62	
Expected egg production		62		47	
Index of Population Trend (I)		$\frac{62 \times 100}{174} = 36\%$		$\frac{47 \times 100}{2176} = 2\%$	

[1] Morris, R. F. and Miller, C. A. (1954) *Canad. J. Zool.*, **32**, 283–301.

* All numbers represent the mean population of 1 m^2 of branch surface.

† This represents the loss of potential reproductive power owing to the sex ratio differing from 50:50.

‡ This represents the loss of potential reproductive power owing to the females being small, and laying fewer than the full complement of eggs.

1.4.3 Population fluctuations

Pests are particularly likely to develop in simple communities, and a characteristic of simple communities is the instability of (lack of precise control in) its constituent animal populations. Where there is a considerable change in the numbers of a pest organism from generation to generation, it is obviously of great practical importance if this can be forecast, as it produces fluctuations in the supply of commodities or recurring crises for health authorities. Emphasis on fluctuations in the size of animal populations takes us from a study of natural control into the broader one of population dynamics. In practice, it has been found that the measurement of only one or two mortality factors has often made successful prediction of population changes possible. Thus, it has been pointed out that, although population densities may be *affected* by many variables, they may be essentially *determined* by only a few, and these have been called 'key factors'.

In Table 8, an 'Index of Population Trend' (I) has been calculated to give

TABLE 9 KEY FACTORS IN THE POPULATION DYNAMICS OF FIVE AGRICULTURAL CROP PESTS*

Pest	*Critical part of the life cycle*	*Percentage of the variation (variance) in I contributed by this stage*	*Degree of density dependence*	*Nature of the key factor*
Fruit tree leaf roller *Archips argyrospilus*	Proportion of potential egg supply actually laid	70	Density dependent	Migration of adults (?)
Diamondback moth *Plutella maculipennis*	Proportion of potential egg supply actually laid	74	Marginal	Weather
European corn borer *Ostrinia nubilalis*	Proportion of potential egg supply actually laid	78	Marginal	Migration of adults (?)
Eye-spotted bud moth *Spilonota ocellana*	Survival of winter larvae	68	Density independent	Frost
Pistil casebearer *Coleophora serratella*	Survival of winter larvae	91	Density dependent	Parasitism by *Epilampsis laricinellae*, and bird predation

* Watt, K. E. F. (1963) Article in 'Population dynamics of agricultural and forest insects'. Ed. Leroux *et al. Mem. Ent. Soc. Canad.* **32**, 83–91.

an overall impression of the numerical performance of the generation in question by comparing the size of one generation at its beginning with that of the next. It is derived as follows:

$$\frac{\text{The number of eggs laid by generation 1}}{\text{The number of eggs laid by generation 2}} \times 100$$

In a species such as *C. fumiferana* with only one generation per year it provides a convenient measure of population variability from year to year. The availability of life tables over a number of years and perhaps in a number of places makes it possible to detect which mortality factors fluctuate in a similar way to fluctuations in I. These must be the key factors, and Table 9 presents an analysis of this sort relating to five pests of agricultural crops. As can be seen, the key factors vary from pest to pest, and it seems reasonable to assume that they will also vary from area to area for any given pest.

1.4.4 Population control by man

Pest populations may not always be under natural control and their consequent fluctuations produce variability in the availability and price of commodities. Even where precise natural population control does exist, the equilibrium density may still be too high, particularly in respect of medical pests, and where from an ecological point of view, man is making ever increasing demands from his food organisms (5–7·5 tonnes of seed from an ha of wheat, 10–13 000 litres of milk per year from a cow, or 260 eggs per year from a hen). Stable, low pest populations are required, and to achieve these man adds new mortality factors, most commonly, introduced predators and parasites of the pest or toxic chemicals. Here the concept of Density Dependent Factors becomes important, because if a population is being naturally controlled at a particular level, and a new source of mortality is added, the Density Dependent Factors originally responsible for control or regulation will now operate less strongly as numbers decline, so producing a measure of compensation.

This is well illustrated in experiments on the sheep blowfly (*Lucilia cuprina*) in Australia by Nicholson (1954)*. He kept 4 blowfly populations on a constant food supply, left one as a control, and from the other three he regularly removed 75, 95 and 99% respectively of the newly emerged adults. Given a finite amount of larval food, the surplus larvae starved, and food which had been eaten by a starving larva was not of course available to the rest—competition for food was acting as a Density Dependent Factor, causing greater mortality at higher population levels. The relaxation of this and other Density Dependent Factors, as a result of Nicholson's imposed mortality produced such efficient compensation in this rather exceptional case, that the number of blowfly adults produced actually increased as a result of his

* Nicholson, A. J. (1954) *Aust. J. Zool.* **2**, 1–8.

predation, 157 adults emerging daily from the control, 192 at 75% predation, 437 at 95% predation, and 909 at 99% predation. Similarly biomass yield in fisheries increases with exploitation up to a critical level after which the animal population is damaged, and extinction may follow. That is to say that compensation for man-imposed mortality is at the expense of natural mortality factors up to a certain point; after that, no further compensation is possible. Whilst this may be of some consolation for fishermen, it provides no comfort for those engaged in pest control as it means that very high percentage kills must be achieved if the final population is to be reduced by a substantial amount for any length of time. The problem of reducing damage by population reduction is an even more depressing situation for in addition to population compensation, account must also be taken of a possible logarithmic relationship, previously mentioned, between populations and damage. Here, only at the very highest population reductions do we get a commensurate reduction in damage.

The effectiveness of toxic chemical treatment is often measured by the percentage of the pests killed, and by this is meant the percentage of visible pests at the time of treatment. In practice, a population survey of the pest is conducted before treatment with the chemical, and after a suitable lapse of time for the poison to exert its effect, a post treatment survey is undertaken. A simple calculation on these data gives the percentage kill. Using such criteria, Webb *et al.* (1961)* estimated that 88% was a common larval kill obtained after the aerial spraying of DDT against spruce budworm in Canada. This is a high figure, and there was commonly a satisfactory saving of 30% of the foliage that would have been destroyed. However, when the more theoretical aspects of pest control are the primary consideration, this 'apparent' percentage mortality (of prime practical importance) should be clearly distinguished from the 'real' mortality. Thus in the second life table in Table 8, compiled for an outbreak area, DDT from just such a control campaign in fact killed 8·3 larvae per 1 m^2 of branch surface. With an overall generation mortality of 99·97%, DDT was responsible for a real mortality of only 0·38%. More significantly, the proportion of the pest population killed by chemical poisons is unlikely to increase as the size of the pest population increases, if only because coverage by the spray is usually incomplete, so that a certain area will be left unpoisoned. Accordingly, poisoning is likely to be density independent in its action. When chemical control measures precede important Density Dependent Factors therefore, compensation is likely to occur; when they come afterwards, a maximum increase in the generation mortality is likely to result. The possibility, as in Nicholson's experiments, of actually increasing the final population size, by introducing early mortality factors which operate before the regulatory ones, is disturbing, and it is also easy to see how the introduction of a toxic chemical which is unlikely to have a regulatory influence can be disastrous if it also kills other

* Webb, F. E., Blais, J. R. and Nash, R. W. (1961) *Canad. Ent.* **93**, 360–379.

biotic agents (parasites, predators and competitors) responsible for important density dependent portions of the total mortality.

Much of the foregoing discussion, of necessity grossly simplified, must still appear to be highly theoretical and of marginal value in practical pest control. Nevertheless it may make some of the apparently anomalous results in practical control programmes more comprehensible. It would clearly be desirable if handy formulae could be produced which would indicate whether or not the spraying of crops or the dosing of cattle would give satisfactory ecological results in particular situations. It would be even better if they could indicate the best chemical to use, the best time to take action, and the optimum dose, but unfortunately our knowledge of the processes involved in population dynamics is still inadequate, and practical recommendations remain largely empirical. Indeed percentage mortalities, and percentage damage or population reduction, may be of less practical relevance than a measure of the time taken for the reduced population to return to its former level, a powerful factor in considering the relative merits of chemical and biological control. Of even greater practical relevance is the cost:potential benefit ratio, provided it is calculated on a long-term basis. The great lesson of soil erosion is that short-term profitability and long-term profitability may be quite different things; this applies with equal force to the measures taken to control pest animals

Chapter Two
CONTROL BY THE PREVENTION OF POPULATIONS FROM ACHIEVING PEST STATUS

It has frequently been emphasised that the old adage, 'Prevention is better than cure', is a valuable guide to all who are concerned in the control of pest animals. In some ways techniques which prevent the build-up of pest damage to serious economic levels are 'control' techniques in the truest sense of the word, and these are the subject of this Chapter.

2.1 Control by the Prevention of Physical Access

One of the ways in which the problem of controlling pest animals can be approached is to prevent the pest and host from contacting each other, the most obvious way of achieving this being to erect some sort of barrier between them. Broadly speaking these barriers can be considered under four headings, legislative, physical, chemical and behavioural.

2.1.1 Legislative barriers

The aim of this type of control is to restrict the spread of pests into new countries, and also, in the case of pests already established, into new territories, whether it be on a local or more widespread scale. This is a most important aspect of pest control since the introduction of animals into new environments, where natural enemies are absent, can result in previously harmless animals becoming destructive pests. In fact many of the examples of the successful use of biological control have been applied to this category of pests (see Chapter 6).

Quarantine regulations are assuming much greater importance in view of the increased speed and amount of travel between widely separated areas of the world. Thus the natural barriers to pest dispersion are no longer adequate in their control. Regulations are provided not only to control the movement of plants and animals but also the growing of certain types of crops, the size of the fields on which they shall be grown, and to enforce the compulsory treatment of certain pests together with other hygiene regulations. Many of the regulations designed to enforce certain cultural practices are aimed primarily at preventing pest infestations. (See Plate 3B.)

Pests are readily transported in seeds, food, livestock including domestic animals, wildlife and even man. "The Importation of Plants Order 1958–59", restricts the introduction of certain plants into Britain unless accompanied by a health certificate issued by the country of origin. Some plants and crops are completely prevented from entry, thus potato tubers grown in countries where the Colorado beetle (*Leptinotarsa decemlineata*) is prevalent may be refused an entry permit. Other plants are only imported from certain countries at seasons when certain pests are least abundant. The main aims of these import restrictions on plants are firstly to prevent the introduction of virus and other plant diseases, and secondly to control the entry of the Colorado beetle, the cherry fruit fly (*Rhagoletis cingulata*), the San José scale (*Quadraspidiotus perniciosus*) and various lepidopteran pests (see Fig. 4).

Particular care is exercised in the import of nursery stock where the chances of the pest being introduced on the plant are extremely high. Regulations also apply to the export of crops and plants. While the above examples of legislative control apply mainly to Britain, similar regulations are applied in other countries and often also between regions or states within one country.

The import and export of livestock, meat and domestic animals, is likewise regulated by rigid legislative measures to control the spread of infection. In order to prevent epidemic outbreaks many highly infectious diseases are notifiable by law and specific orders are available to deal with them. In Britain, sheep scab and parasitic mange of sheep and horses are examples of infectious diseases which fall within the scope of this book.

Meat for human consumption must be officially inspected before it can be sold. Other regulations exist to control, for example, the disposal of sewage and dead livestock, and the treatment of warble fly in cattle, but unfortunately such regulations are too often ignored. It has already been mentioned that many cultural procedures are controlled by legislation, the sugar beet orders of 1960 and 1962 being good examples of this in the United Kingdom. The plant parasitic nematode *Heterodera schachtii* (sugar beet eelworm) is an exceptionally destructive pest attacking the root system, thereby causing wilt which often results in the complete destruction of the host plant. Crop rotation of infested land is enforced to ensure that susceptible crops are not grown in some years thereby preventing the build-up of the population of this pest in the soil. In certain scheduled areas on all land whether infested or clean, a compulsory three-course rotation is demanded, while in cases of moderate infestations a four-course rotation is enforced. In very heavily infested areas no susceptible crop is grown for a period of at least five years and only then after the land has been shown to be free from this particular pest. The spread and increase of other cyst nematodes, for example the potato root eelworm (*Heterodera rostochiensis*) (see Fig. 9 and Plate 2B), is also rigidly controlled by regulatory measures of a domestic and international nature.

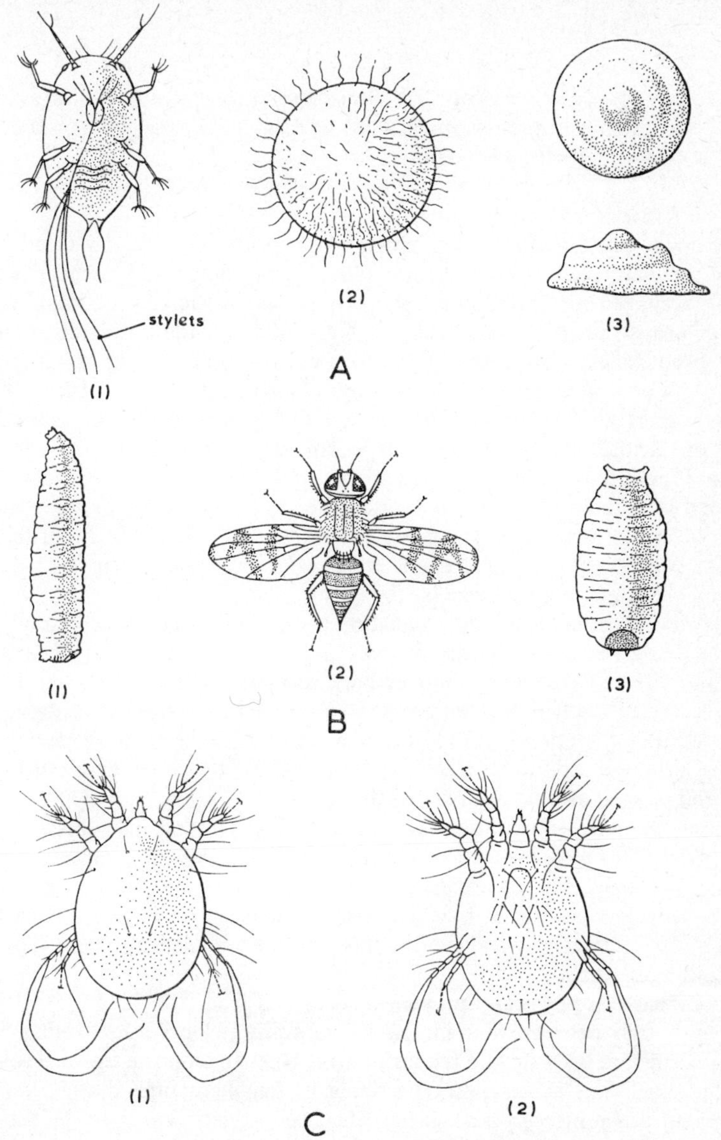

Fig. 4 Some important pests against which quarantine regulations are directed.
A. San José scale, *Quadraspidiotus perniciosus*
(1) Ventral view of larva showing the stylets which pierce the plant and conduct its sap into the insect. (2) Dorsal view of later stage in larval life showing protective waxy filaments developing. (3) Dorsal and lateral views of fully formed young scale.
B. The Cherry fruit-fly, *Rhagoletis cingulata*
(1) Maggot. (2) Fly. (3) Puparium.
C. *Psoroptes ovis* the causative organism of sheep scab.
(1) Dorsal view. (2) Ventral view.
Diagrams not drawn to scale.

The control of sugar beet eelworm illustrates how cultural and legislative control are often completely integrated, and how difficult it is to treat in isolation different methods of control.

2.1.2 Physical barriers

Barriers of this type physically prevent the contact between host and pest. Firstly we shall refer to those barriers which are designed to prevent the pest gaining access to its host. Fences are often used in the case of mammalian pests of crops. Thus in Australia and New Zealand, thousands of miles of rabbit-proof fences have been erected to prevent rabbits damaging crops and pastures. These animals also destroy woodland by grazing on young trees and by damaging the bark of older ones. The cost of providing protective fences in British woodland alone is estimated to exceed £2 000 000 per annum. The intense grazing of pasture by rabbits encourages the growth of unpalatable vegetation or can so denude the vegetation cover that secondary effects of erosion may result. After the myxomatosis epidemic in Britain in 1953–54 (see Plate 7A) large areas of grazing land were seen to improve due to the removal of grazing pressure by this pest.

Deer, too, damage woodland, by eating the shoots and buds of young trees and removing the bark of older ones. The red deer eats the bark but it, in common with the roe deer, removes bark to mark territory. Damage to the bark also results when the antlers are rubbed against trees to remove the 'velvet'. Full protection can only be provided by the erection of deer-proof fencing although individual trees can be protected by either wire netting or metal foil twisted around the base of the tree.

Among the smaller mammals rats and mice are important pests of foodstuffs as well as presenting a considerable health hazard to man and his domestic animals. The provision of adequately rat-proofed buildings would prevent the damage caused by this pest. Methods of achieving this include the use of concrete or metal food stores, and of mechanical barriers placed in the animals' runs.

It is difficult to physically prevent insect pests from attacking animals and crops although nets prevent attack by mosquitoes and other biting flies. Sticky bands are used on the trunks of fruit trees to stop the upward migration of non-flying insect pests particularly female winter moths, various weevil and other insect pests. Other physical barriers are used against termites, which are widely distributed in the tropics and sub-tropics where they cause damage to timber. This damage to wooden buildings can be minimised by the construction of concrete foundations provided with a two inch projecting shelf both inside and outside the building. This prevents the termites from constructing their runs into the wooden parts above, since this pest has difficulty in constructing the run when working in an inverted position. Caps placed on top of masonry pillars which support wooden structures achieve the same purpose (see Fig. 5).

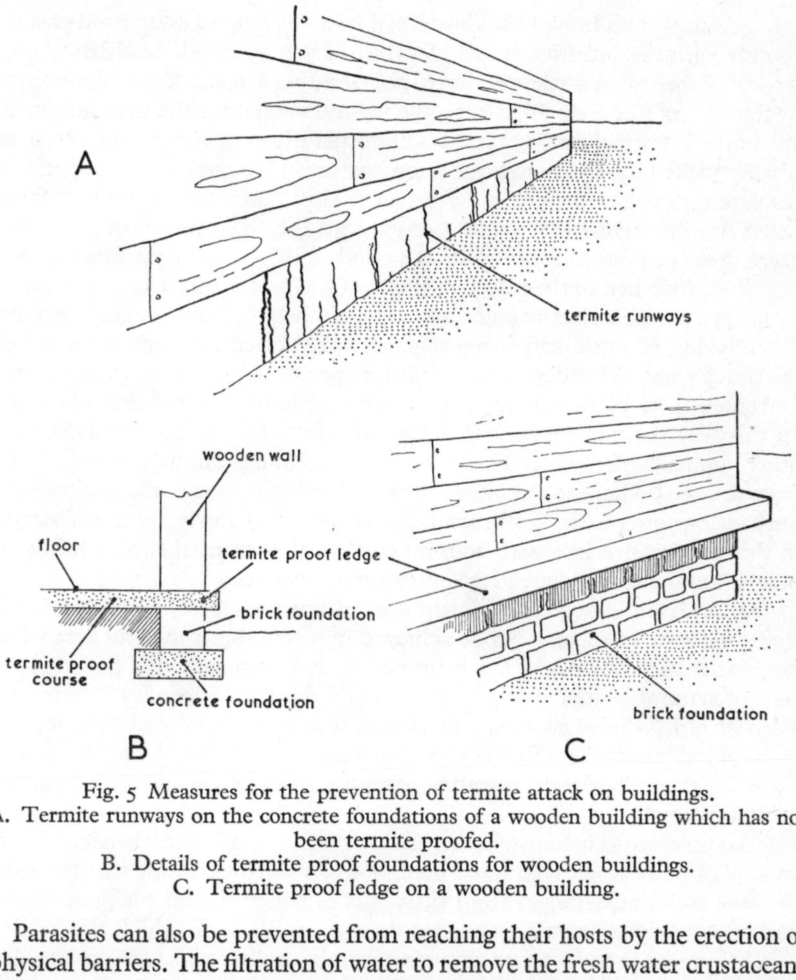

Fig. 5 Measures for the prevention of termite attack on buildings.
A. Termite runways on the concrete foundations of a wooden building which has not been termite proofed.
B. Details of termite proof foundations for wooden buildings.
C. Termite proof ledge on a wooden building.

Parasites can also be prevented from reaching their hosts by the erection of physical barriers. The filtration of water to remove the fresh water crustaceans *Cyclops* spp. which act as the intermediate host of the guinea worm *Dracunculus medinensis* of man (see Plate 6C) is an example of this. Hookworms infect man when their larvae penetrate the skin of unprotected parts of the body, particularly the feet. Its incidence could be reduced if shoes were worn. Schistosomes, important blood dwelling trematode parasites of man, can be prevented from establishing themselves if care is taken to ensure that parts of the body likely to come in contact with water containing the infective stages of the parasites (cercariae) are adequately covered (see Plate 4).

An alternative approach to the prevention of contact between host and pest is to keep the host away from the pest. An example of this is seen in the control of intestinal and other parasites of farm stock. By using the system of zero grazing (grass brought to the animal which is housed away from pasture) contact with the infective stages of parasites is minimised. Similarly the infection of sheep and cattle with liverfluke (*Fasciola hepatica*), can be prevented by fencing off wet areas of pasture, the natural habitat for the snail intermediate host. Intestinal and other nematode parasites of sheep and cattle are usually picked up by these hosts from pasture. However, in most species of nematode parasites there is a delay before eggs deposited in the host faeces reach the infective larval stage. Consequently, by allowing host animals to graze areas of pasture for only short periods of time, and then moving them to a fresh strip before the infective stages of the parasite develop, for example by using a portable electric fence, contact with these infective stages is avoided. The grazing of cattle and sheep together on infected pasture may also help to reduce parasite burdens since parasites specific to sheep, for example, may be removed by cattle (in which they are unable to develop) and *vice versa.* In this way the one animal can act as an effective 'vacuum cleaner' for the other animal's parasites. The running of immune animals over infected pasture can also help to clean the herbage of infective stages of parasites without causing injury to the host animal (see section 2.3). Clearly, the use of barriers in this context overlaps with cultural control, the essential barrier being one of space and entirely dependent on cultural practices.

The problem of segregating field crops from their pests is more difficult. Some measure of success can be achieved by siting the crop in an area where the particular pests to which it is prone are least abundant, and this exploitation of natural spatial barriers in crop and animal husbandry is often the purpose of legislative control. The compulsory growing of seed potatoes only in specified areas where the incidence of virus transmission by aphids is very low is an example of the integration of cultural, legislative and spatial barriers in pest control.

A complete discussion of the relative merits of physical barriers in the control of pests is not warranted here, but it is worth stressing that the value of these techniques ranges from situations in which the host is permanently and completely protected from pest damage, to those in which the barriers are little more than stop-gap measures, rarely dealing with the real problem of controlling the pests' population. While many of these methods of control are of little more than academic importance some are of considerable use in preventing pest damage and are relatively inexpensive. The control of liverfluke, for example, by fencing off areas of wet pasture can be completely successful without involving the farmer in a great deal of expense.

2.1.3 Chemical barriers

Chemical pesticides are often used in the prevention of pest attack. Wide-scale spraying of crops is a normal procedure in the control of some pests. The chemical is used to prevent the pest from invading the host by erecting a chemical barrier which is lethal to the pests. This is rather different from the usual use of pesticides considered in Chapters 3 and 4, for the direct reduction of pest populations present in outbreak numbers. This preventative spraying, as it is sometimes called, is standard agricultural practice in the growing of certain crops which are particularly susceptible to insect and fungus attack. Dieldrin and aldrin (see section 4.2.4) are used in the prevention of termite attack in buildings, these chemicals being placed in the soil, surrounding the foundations. With preventative spraying it is important that the activity of the pesticide persists for a long time. The obvious disadvantages of this in terms of resistance, are discussed in 8.2.1. and also in terms of hazards due to residues in 8.1.

2.1.4 Behavioural barriers

Pests locate their hosts by reacting to external stimuli such as sight, smell, touch and taste, and it is therefore possible to protect the host by exploiting the normal behavioural responses. This can be achieved by means of frightening devices, or providing attractive alternatives to the host, or by rendering the host unattractive. For example, attractants mixed with pesticides can be used to attract the pest away from its host while chemical and other repellents actively dissuade the pest from attacking.

(a) ATTRACTANTS

The use of attractants to increase the efficiency of control measures has been extensively examined in recent years. Their main advantage is that it is unnecessary to locate the pest but instead the pest comes to the lure. The most potent of the insect attractants so far investigated are effective over a distance of 4 km. They are not only highly potent but also very specific, attracting single species or closely related insects and often the male sex only. Such attractants have already proved useful in the early detection of pest infestations and consequently also in the forecasting of pest attack. The Mediterranean fruit fly (*Ceratitis capitata*), an important pest of peaches, apricots, nectarines, oranges and prickly pears in many parts of the world, was successfully eradicated from Florida in 1957. Attractants were used to indicate reinfestations in areas already treated with insecticides, so that flies surviving the first application could be detected and resprayed.

Attractants are useful too in the direct control of insect pests. Thus by combining an attractant with an insecticide in a bait, more specific control can be achieved than by blanket spraying with the insecticide alone. Furthermore the complete treatment of the infested area is not necessary since the pest can be lured to the pesticide. In a similar manner effective control has been achieved using attractants combined with chemosterilants (see section 6.2). The use of attractants in conjunction with pesticides also has the advantage that broad spectrum pesticides which may be harmful to useful species can be employed by placing the specific attractant/pesticide mixture in positions in the field which are accessible only to the pest.

Attractants can be classified either on the basis of their function, or on the source from which they may be derived. Functionally, there are various types of attractants which bring animals together for various purposes such as mating (i.e. sexual attractants), feeding, oviposition, defence, and hibernation. There are two natural sources of insect attractants; the host plant or animal may produce chemicals which stimulate the taste or olfactory senses of the insect, or the insect itself may produce chemicals known as pheromones, which amongst other things serve to bring together individuals of the same species, again by stimulating the chemo-sensory receptors (see Plate 3A).

(i) Sexual attractants. Many insects produce pheromones for mating purposes and these are detected in incredibly small amounts by the opposite sex. Males of the American cockroach, *Periplaneta americana* for example, can detect 10^{-14} microgrammes (or 30 molecules) of the female pheromone. The gypsy moth *Porthetria dispar* can be trapped from distances of two miles using specific attractants. The existence of these compounds is probably quite widespread among insects and it has been postulated that their evolution has frequently occurred in those species which have ceased to feed in the adult stage (the above examples being exceptions). In these species the sexes obviously no longer congregate on food material, and some other mechanism becomes essential to ensure a viable frequency of meetings. The potency of these compounds can be quite considerable, a single female attracting large numbers of male individuals within a few hours. The majority of sex attractants are species specific although some can attract closely related species. It has been shown that the gypsy moth pheromone is 10-acetoxy-cis-7 hexadecen-1-ol which is a relatively simple substituted straight chain aliphatic alcohol. A closely related compound to this has been synthesised in the U.S.A. and is marketed under the trade name of gyplure, where it has been extensively used for estimating the density and distribution of this pest in the field. The structure of the sexual attractant isolated from the female silk moth (*Bombyx mori*) resembles that of gyplure, being a straight-chain aliphatic alcohol although unsubstituted = trans-10 cis 12 hexadecadien-1-ol, and it seems likely that other lepidopterans may have chemically related attractants. Sexual attractants have also been used to assess the incidence of the tobacco hornworm moth (*Protoparce sexta*), the pink cotton bollworm moth (*Pectinophora gossypiella*) and the pine sawfly (*Diprion pini*).

(ii) Food and ovipositional attractants. Many species of insects are attracted to their hosts by olfactory stimuli, and these may be due either to chemicals produced by the host or to pheromones deposited during previous visits to the host by other individuals of the same pest species. Information on this aspect of pheromone function is accumulating rapidly at the present time, good examples being the substances used in scent trails by ants and termites, and those deposited by houseflies on their food which renders it much more susceptible to subsequent fly attack. It is probable that pheromones of this type will be exploited in future developments in insect control, but at the present time far more attention has been concentrated upon other naturally occurring or synthetic attractants. Thus animal perspiration, blood and urine are all attractive to mosquitoes, but little information is available on the precise chemical nature of many of these natural attractants which direct pests to their prey. Generally it is found that many species of Diptera oviposit in the vicinity of ammonia, amines, sulphides and fatty acids, probably because these chemicals are normally associated with decomposing organic material, a source of food for the larval stages of these insects. Protein hydrolyzates used in conjunction with insecticides have proved successful in the control of the Queensland fruit fly (*Stutmeta tryoni*) in New South Wales. The protein hydrolyzate plus insecticide when sprayed on to foliage attracts both sexes of the pest. A derivative of phenyl butanone, again mixed with an insecticide, has also been used successfully as an attractant for males of this pest.

The search for attractants among foodstuffs has not been particularly rewarding although a few have been isolated from purified plant extracts. For example methyl eugenol from the huon pine, and citronella oil have both proved effective as lures for the oriental fruit fly (*Dacus dorsalis*). Unfortunately, however, many of these extracts lack specificity and furthermore do not retain their activity for long periods of time. Attention is now turning towards synthetic chemicals in the search for more powerful and specific attractants. The synthetic attractants 'siglure' 'medlure' and 'trimedlure' are very active for the Mediterranean fruit fly (*Ceratitis capitata*) in spite of the fact that they do not, as far as is known, occur in nature. The reason why insects should be attracted to such compounds remains a mystery.

The use of attractants has many advantages over more conventional methods of pest control. Not least of these is the use of attractants with pesticides which reduces the likelihood of toxic hazards and also increases the ease of locating the particular pest. An additional advantage is that the pesticide need not be applied directly to the crop, an obvious factor in preventing side effects due to residues. It is conceivable that resistance to food and ovipositional attractants mixed with insecticides could result from selection of unresponsive individuals in the surviving population. This is less likely to occur in the case of attractants since selection for unresponsive individuals would be a decided disadvantage in perpetuating the pest population.

(b) REPELLENTS

In a similar way to attractants, repellents involve interference with the response of the pest to its host. Repellents can be frightening devices, which are commonly used for example to dissuade birds from attacking crops, or chemical repellents, which are used to make the host objectionable to the pest. The use of scarecrows to frighten birds is an example, albeit an inefficient one, of a frightening device. Herring gulls, rooks and starlings often leave their feeding and roosting sites if recordings of the distress calls of their own species are played. Wood pigeons respond to white bird-shaped objects flown overhead by leaving their feeding sites to dive for cover, presumably as a natural response to predatory birds. Explosives, rockets and electric flashes have been used experimentally in the U.S.A. to frighten birds away from economically valuable crops.

Field tests on the use of sound to repel the European corn borer *Ostrinia nubilalis* have been conducted in the U.S.A., and have shown that by simulating the sounds produced by echo-locating bats, a natural predator of this night-flying moth, the incidence of oviposition by the adult moth on maize was more than halved.

Certain species of arthropods produce offensive smelling secretions when disturbed, a defence mechanism intended to dissuade predators from attacking. It would seem possible, therefore, that natural repellents may also be of use in protecting animals and plants from pest attack, and several essential oils obtained from plants have been examined as potential repellents against insects particularly sandflies, midges and mosquitoes. Examples of these oils which repel mosquitoes are those extracted from the huon pine, *Eucalyptus* and *Caryophyllum*. Bark beetles are repelled by vapours of non-host resins and attracted by those of the normal host plants. The volatile fractions obtained from cotton seedlings when painted on to cotton seedlings themselves produced 100% repulsion of adult boll weevil (*Anthonomus grandis*) for five hours while untreated controls were completely destroyed within two hours.

Much of the research into repellents has been conducted against mosquitoes, biting flies, fleas, ticks and chiggers. Dimethyl phthalate, indalone, 2-ethyl-1-3 hexanediol and NN diethyl-m-toluamide are examples of chemicals which have been shown to effectively repel mosquitoes. A very important factor in assessing the efficiency of repellents is the loss and changes in the nature of the repellent itself as a result of both the action of the host and the environment. Thus sweating in man and animals tends to decrease the effectiveness of repellents through its diluting effect. The main factor however, is the absorption of the repellent via the 'skin' of the host, although some also undergo chemical change with time.

DDT has a repellent effect on some insects; certain mosquitoes for example seldom rest on DDT sprayed surfaces and often leave sprayed buildings. The insecticide appears to cause an undirected excitation of the insect after a very short period in contact with the chemical and before it has acquired a lethal dose. This follows from the stimulation of sense organs on the tarsi of these flies and is one of the mechanisms by which behavioural resistance can arise (see section 8.2.1). In the case of stable flies (*Stomoxys* spp.) and the house fly (*Musca domestica*) the stimulus from repellents is detected by the antennae, since removal of these organs prevents them from directing themselves away from repellent vapours. Mosquitoes are normally attracted to a warm moist surface and many mosquito repellents interfere with this orientation reaction. If, however, the insect manages to accidentally alight on the treated host, then secondary contact repulsion dissuades the fly from actually biting. In some cases this repulsion acts through the mouthparts before they penetrate rather than through the tarsi and antennae and mosquitoes have been observed to stand in these repellents and bite through an unprotected part of the host.

Repellents have also been used with limited success against birds and mammals. Anthroquinone, naphthalene and various alums have been advocated as repellents against pest birds but field trials have not been particularly successful. Sticky repellents painted on to the buds of fruit trees have had some success in controlling the activities of bullfinches in orchards. The treatment of seeds of cereals with coal-tar preparations has had some value in reducing the losses due to crows, starlings and other birds in newly planted fields. Mice and voles can cause extensive damage to nursery stock, particularly young trees, and sulphur-linseed oil preparations have been found to produce good repellent action against those pests. Finally it is worth mentioning the use of 'jelly-like' strips of polythene plastic which prevent starlings and pigeons from alighting on buildings in towns, by virtue of the fact that contact of their feet with these surfaces disturbs the normal roosting behaviour of these birds.

2.2 Cultural Control

2.2.1 Introduction

Cultural techniques of pest control involve the manipulation of crop and animal husbandry practices in such a way that damage is minimised, at the same time permitting the production of an economically viable commodity.

In this section we are concerned with those cultural practices which are worth adopting where one has grounds for expecting pest attack. It can readily be appreciated that the assessment of levels of potential damage and, more important, whether it is economically worth while initiating preventive measures is extremely difficult in the absence of accurate objective information. New knowledge of the biological relationships between pests and their hosts will continue to give greater understanding of the complex epidemological picture of pest attack and technical advances will develop to exploit this information. Each advance will require fresh scientific and economic assessments of suitability.

2.2.2 Prediction of pest infestations associated with certain seasons

Some attempts have been, and are being, made to predict the potential incidence of pests and the diseases they cause. The observation that marked numerical fluctuations in pest members are associated with certain seasons or meteorological conditions has led to the development of forecasting systems.

Success has been achieved in forecasting the incidence of such insects as the codling moth of apple (*Laspeyrisia pomonella*), the spruce budworm (*Choristoneura fumiferana*) in Canada, and the desert locust (*Schistocerca gregaria*). Methods have been developed for forecasting the incidence of liver fluke (*Fasciola hepatica*) in cattle and sheep in Britain, and the incidence of nema-

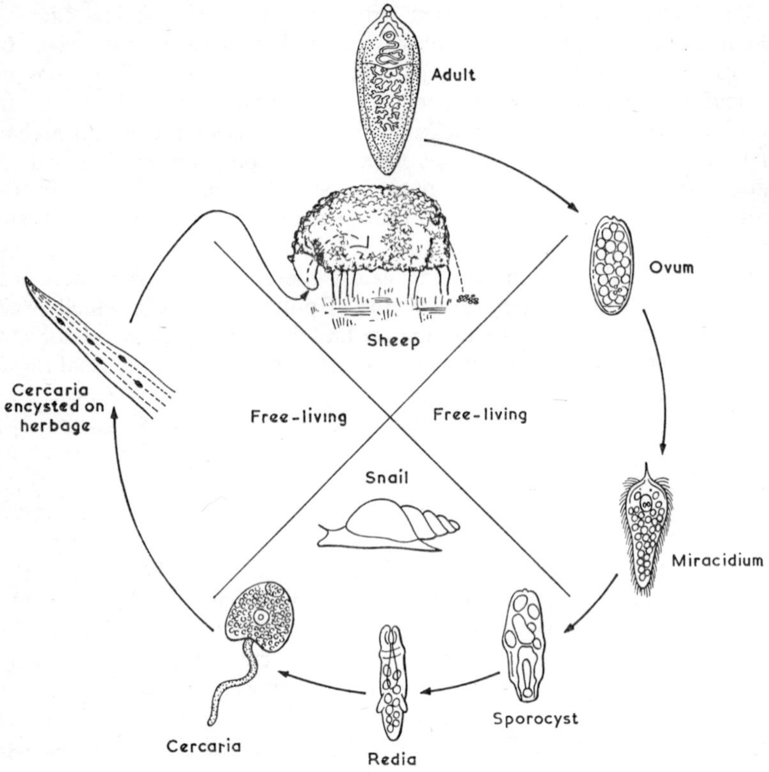

Fig. 6 A. The life cycle of *Fasciola hepatica*, the liver fluke. The adult fluke lays eggs which are passed down the liver bile ducts to the intestines, and thence out in the faeces. The miracidium hatches from the ovum (egg) and swims towards a snail, *Lymnaea truncatula*, penetrates it and there develops into a sporocyst. Each sporocyst produces many redia and each redia produces many cercariae. The cercariae emerge from the snail in warm damp weather and swim in films of moisture before encysting on herbage. The definitive host, sheep, cattle and numerous other animals, then eats this herbage and the young liver flukes migrate from the intestine to the liver capsule and from there to the bile ducts. During this migration they cause considerable damage. If present in sufficient numbers the host may be killed from the haemorrhage caused by these immature migrating flukes (*Acute fluke disease*) before they reach the bile ducts. If the infestation is more gradual then these flukes reach the bile ducts, mature, and eventually produce eggs. Disease caused by adult flukes is called *Chronic fluke disease*.

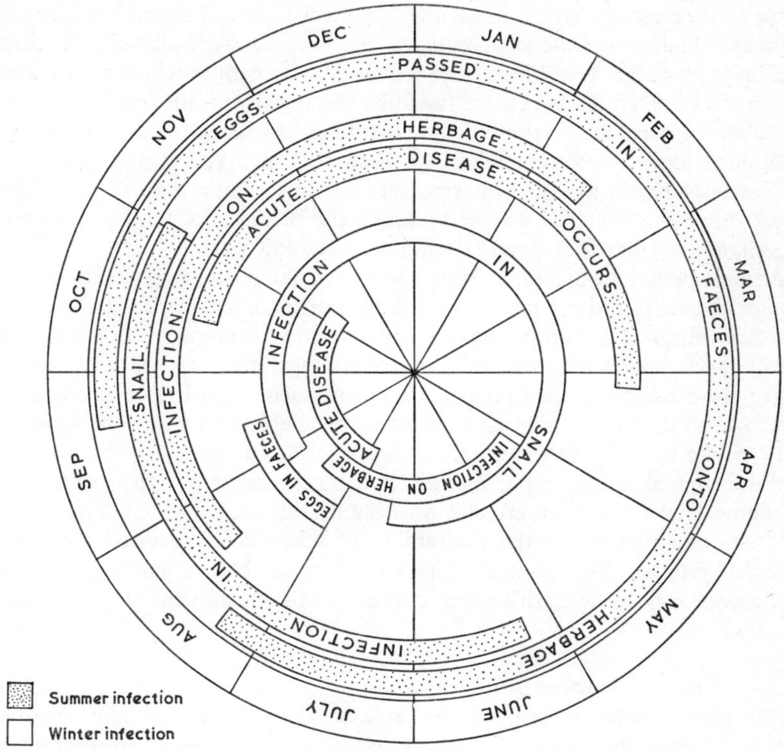

B. The timing of the life cycle of *Fasciola hepatica* in Britain (from Ollerenshaw, C.B. 1959).

todiriasis (*Nematodirus* spp.—important intestinal nematode infestations) of lambs.

In the case of liver fluke the ability of the parasite to survive depends upon the ecological association of the definitive host with the snail intermediate host (*Lymnaea truncatula* in the U.K.). Any environmental factor such as climate which influences the survival and fecundity of this snail will influence the survival of *Fasciola hepatica.* Snails survive from one year to another but only grow and reproduce rapidly in wet conditions where the prevailing temperature remains above 10°C. Similarly, fluke eggs deposited by infested hosts only hatch at temperatures above 9°C and the miracidum (see Fig. 6A) must have water in which to swim and must reach a snail within 24 hours of hatching. Further, the fluke develops more quickly and more infective stages (cercariae) are produced during warm wet conditions (see Fig. 6B). Thus, the forecasting system developed for liver fluke takes account of the habits and density of the snail as affected by climatic factors and the likely production of cercariae (later metacercariae), which in turn will influence the course of the disease in animals which graze the contaminated pasture. As with all effective forecasting systems it is possible to issue warnings as a result of which appropriate action can be taken. In the case of the fluke this usually means keeping animals off 'fluky pasture' at the time of the year when the cercariae are leaving the snail host and encysting on herbage (see section 2.1.2), using a suitable molluscicide to kill the snail intermediate host, vaccinating the sheep against Black disease (a bacterial disease to which the presence of fluke predisposes sheep), and regular therapeutic dosing against adult flukes.

A further example of a forecasting system is used in the control of the desert locust. Swarms of these insects ravage vast areas of Africa, India, Pakistan and Asia Minor. Data on swarm movements and breeding activity is collected in the field along with meteorological recordings, and collated at the Anti-locust Research centre in London. This centre sends out long-term strategic forecasts so that stores and equipment are available and ready for action at appropriate sites. Short term tactical forecasts for guidance of personnel and mobile control units, especially aircraft, necessitated by the continually changing locust situation, are also possible by this means.

These examples show the desirability of scientifically devised long-term forecast systems, although in the final analysis field observations will continue to be necessary to deal with specific local conditions and the pest incidence as it occurs.

2.2.3 The promotion of host vigour

The genetic potential of plants and animals cannot be exploited unless they are maintained in vigorous health. It is therefore necessary to attempt to ensure no, or minimal, contact with pests, and also the ability to withstand or recover as quickly and with as little loss in potential production as possible should pest attack occur.

Before proceeding further it is necessary to explain that the terms infection and infestation are often used interchangeably and this system is adopted here. Infestation and infection thus describe what is basically the same process which follows the presentation of a pest on or within a potential host. It is necessary to distinguish between resistance to infection (natural, chemical or mechanical barriers which inhibit the penetration of host tissues by the pest) and resistance to the development of the disease process. Resistance to infection and disease, and the antithesis, susceptibility, are relative phenomena as Figure 7 shows.

Infection (or infestation) is not, however, synonymous with disease and a whole spectrum of host-pest interaction exists which may result in no or variable symptoms of disease, the symptoms regressing or progressing according to the prevailing state of the relationship. This host-pest relationship is in dynamic balance and alters due to any factor which affects this balance, such as the level of infection suffered, host age, previous host sensitisation

FIG. 7 SPECTRUM OF ANIMAL HOST-PATHOGEN INTERACTION (modified from Schneider, 1946)[1]

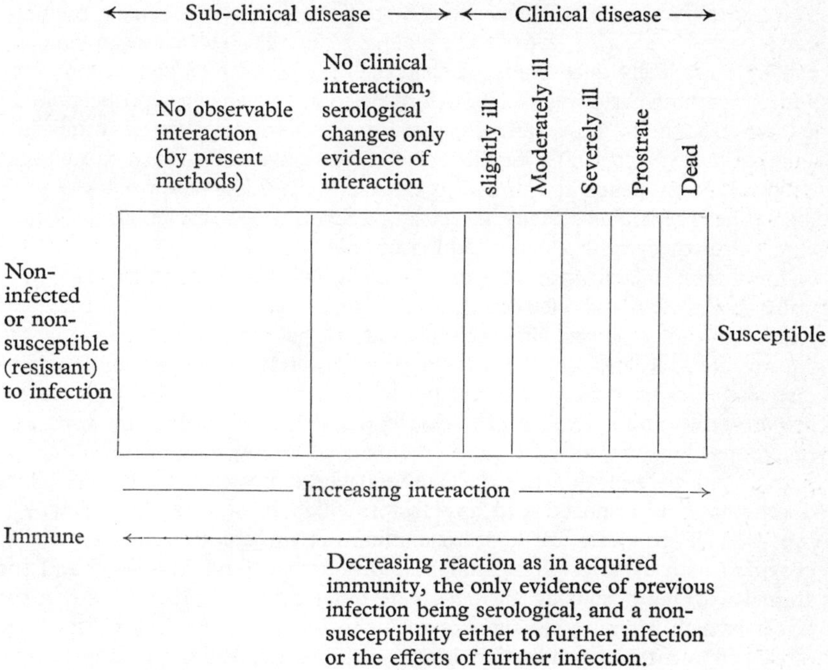

Disease is a relative term which can be applied at any position along this spectrum, and the balance between parasite and host is subject to fluctuations. The disease may be acute followed by rapid recovery or death, or it may be chronic and persist.

[1] Schneider, H. A. 1946. In *Vitamins and hormones*, 4, 35–70.

to the pest, intercurrent infection, diet and state of nutrition, and any other physiological or pathological condition which affects both pest and host.

Figure 7 describes the host-parasite interaction in animals but an analogous situation occurs in the plant-pest relationship, except that the phenomenon of acquired immunity does not appear to exist in plants (see also 2.3).

Each stage of animal or plant development is accompanied by a characteristic range of pests and diseases with which they are particularly likely to be afflicted. Thus the first generation frit fly (*Oscinella frit*) usually invades the young oat plant (see p. 53); ascarid worm infestations are more common in puppies than in older dogs. Generally it is the young animal or plant that is most susceptible to infection and least resistant to the effects of pest damage.

The young animal has not, in general, acquired immunity to the vast range of pests to which it is liable to be exposed during its life and may not have the capacity to develop an effective immune response, especially during the neonatal period (see also p. 63). The young plant, like the young animal, is usually more likely to be damaged than the older plant by a given number of pests since at an early stage of growth the amount of damage is proportionally greater relative to host size. Furthermore the young plant is often less chemically and physically resilient and there has been no chance to build up reserves on which it can call to replace damaged or lost tissues or synthesise pest repellents. Many insect and eelworm pests preferentially attack young growth even of established mature plants.

The permitted life span of a domesticated animal or crop plant is determined by its yield and once this declines below an acceptable limit it is culled. Thus, such animals or crops are culled at a time which is well short of the natural life span and so the husbandryman is rarely faced with problems of disease which are characteristic of old age. Indeed, early maturity is sought in both crops and animals and this accentuates the importance of disease in the young. Early maturity implies an enhanced growth rate of the animal or plant involved and this can only follow where any factor which inhibits development is removed and any factor which is necessary to growth is provided. Thus pests are kept to a minimum or eradicated, the plant is provided with optimal physical soil conditions and food (fertiliser) and the animal with good housing and adequate nutrients, bearing in mind that the factors which promote vigour may vary considerably according to the stage of development and production involved. Thus the young plant or animal has a higher nitrate or protein requirement respectively for optimal growth compared with the mature stage of development. The pregnant or lactating animal has a different dietary requirement from the barren animal. Similarly, the diseased animal or plant is liable to have a different nutrient requirement from that of the healthy and may also benefit from the provision of a different physical environment. Thus, when considering host vigour one needs to be aware of how to maintain vigour in the healthy and how to retrieve it in the diseased animal or plant.

Both deficiency of some essential factor or excess of another creates imbalance in a host and this will in turn be reflected in the host-parasite relationship. Epidemiological surveys show that severe outbreaks of disease in man and domestic animals often accompany periods of famine and economic and other social stresses. The influence of undernutrition and malnutrition affecting the course of infection are particularly well seen in areas where malaria and hookworm infestations are common in man and where gastrointestinal parasitism by nematodes is found in domesticated ruminants. In all instances therapy needs to include the provision of an adequate diet; indeed, there are records of rapid cure of a number of disease conditions in animals that have simply been provided with such a diet. An example of how diet can influence the degree of infection as well as the course of disease following infection is provided by the lactating sheep and her young. The ewe kept on an inadequate diet cannot provide sufficient milk for her offspring (especially if twins or triplets are born) to satisfy its hunger and requirements for growth. Such a lamb may turn to grass to supplement its food and consequently pick up nematode parasites in the herbage. If the animal is already lowered in vigour the additional stress of parasitic disease may cause death or irreparable damage. Healthy ewes with a good milk yield satisfy the appetite and nutrient demands of the young lamb so that when the lambs do start to eat grass they are sufficiently advanced in development that they can tolerate parasite infestations more readily.

Dietary deficiencies, both qualitative and quantitative, are very commonly seen in all animals and plants. Such conditions are readily seen in man in the underdeveloped parts of the world, for example where carbohydrate rich material is the staple diet and protein deficiency is common. Kwashiorkor, the term given to describe the body condition caused by protein deficiency, is commonly aggravated by enteric amoebae and worms which interfere with the host's ability to absorb the limited amount of protein contained in the diet. Indeed, it is a striking fact that deficiency or 'stunting' diseases in both animals and plants may be directly attributable to the activities of pest animals. (See Plates 6A and 6B.)

The dietary intake of an animal or plant may have a direct effect upon the pest. Evidence has been produced which indicates that carbohydrate rich diets favour the survival and vigour of certain intestinal dwelling parasites, especially tapeworms and amoebae, and the converse is apparently true of protein rich diets.

Perhaps the most important influence of nutrition on infestations is indirect, in its influence on resistance mechanisms and the repair of damage. A diet which is sufficient in amino acids, minerals, vitamins and other essential materials provides the host with those constituents concerned in the maintenance of cell integrity and antibody production. Platt (1953)* observed that in an area of Gambia where malaria was endemic the reduction in malaria in one village and improved nutrition in another improved health in strikingly similar fashion.

A critical assessment of the importance of any particular constituent of diet in aiding plants and animals to withstand pest attack is extremely difficult since it is almost impossible, especially in animals, to separate the multitude of influences that have a bearing on host parasite balance. Much of our information stems from observations on deficiency or excess of a specific nutrient or nutrients upon the course of a known experimental infection. The situation can be summarised as follows: pests appropriate nutrients from the host's food tissues, the metabolism of the host is impaired and there is

* Platt, B. S. 1953. Cited in Chandler, A. C. Interrelations between nutrition and infectious disease in the tropics. *Am. J. Trop. Med. Hyg.* **6**, 195–208.

frequently increased catabolism of proteins and decreased utilisation of food. Pests living on or in the host's asorptive surfaces (the root and/or leaf in the case of plants and the gut in the case of animals) impair absorption, or digestion and absorption respectively. In animals appetite and hence nutrient intake is often significantly depressed. Pregnancy and lactation, amongst a number of physiological conditions, affect susceptibility to disease. Latent malaria may flare up in a pregnant woman, and the fecundity of a population of gut-dwelling nematodes may increase in pregnant and lactating animals. The increased numbers of eggs or larvae voided in the dung of the parturient animal increase the risk to the offspring which pick up the infective stages of the parasites.

Studies in the U.S.A. have shown that 'stress' induced by moving animals vast distances from one part of the country to another without adequate rest and food may so reduce their condition that tremendous losses may occur. It has been shown that the host-parasite balance may be shifted so dramatically that parasitic worms which had previously caused little effect eventually killed the host.

The effects of fighting or the prolonged effect of its tensions upon susceptibility to disease is well documented for a number of disorders of man. Comparatively little study has been made of the relationships of animal pest infestations in domesticated animals and men suffering such stresses. It has been found that rats kept in a state of tension with other rats, with which they naturally fight, show a significantly increased susceptibility to *Trichinella spiralis* infection compared with animals that live in harmony.

Parasite infestation may reduce mental as well as physical vigour. Thus nematodes which invade the brain during part of their development may give rise to nervous symptoms: it is considered that nematodes in the intestinal tract may induce similar effects and a number of records of this phenomenon exist from studies on experimental laboratory animals. Thus, it has been found that low grade infection with *Schistosoma mansoni* retarded the learning process of albino rats during a four week trial with a "T" maze, the retardation being proportional to the intensity of infection.

2.2.4 Hygiene

The importance of hygiene for human populations is obvious and it is no less so for domestic animals and plants. This has been recognised since early times and the laws of Moses, documented in The Bible, were formulated as a code of living to ensure, so far as was possible, that the 'Children of Israel' and their animals should remain free of the plagues that regularly afflicted the people of the Middle East and their sources of food. These laws were based on observations of the natural recurrence of certain diseases, in the complete absence of knowledge about what disease agents were involved or how they exerted their effects.

Today, the study of preventive medicine assumes enormous importance and is responsible for the development and maintenance of the elaborate machinery of legislation, hygienic practices, and special procedures of immunisation, vaccination, and other measures designed to ensure that man and animals remain either free of pathogens, or that if they contract them, that they should be able to withstand their virulence.

The hygienic measures adopted in any particular situation are largely dependent upon economic factors, social custom, and an increased awareness of their importance at the personal and communal level. Most children in temperate regions of the world suffer small burdens of the pin-worm (*Enterobius vermicularis*) at some time in their lives, and these depend for their perpetuation on the fact that small children often eat dirt and suck fingers or other objects that may be contaminated with the eggs of this worm. Schistosomiasis, a general term referring to a complex of disease conditions caused by various species of blood flukes, afflicts millions of people in large areas of Africa, the Middle East, Asia, Central and South America, and exists because the people of these countries drink, bathe, or work in water that contains the infective stage of the fluke. Access to the host is by boring their way through the skin or mucosal surfaces (see Plate 4). The parasites are present in the water because human excrement is deposited in irrigation channels and other waterways. In some parts of the world human excrement is a valued, and often the only organic fertilizer and if it does not receive adequate treatment before use it is a potential hazard to all who come in contact with it because of the multitude of pathogens it may contain.

The diet itself may be the source of certain pests; thus the practice of eating undercooked meats and fish has perpetuated the pork (*Taenia solium*), beef (*T. saginata*), and broad fish (*Diphyllobothrium latum*) tapeworms, as well as other worms such as *Trichinella spiralis*. *T. spiralis* remains an important cause of death and illness in man in the U.S.A., Canada, Mexico, Chile, parts of Europe and the U.S.S.R. It occurs in outbreaks involving one or a few families which have eaten raw pork, often as smoked sausage, which is infected with larvae. The latter develop in the intestine and on maturity the female worm transmits a new generation of larvae into the intestinal blood vessels and thence they pass to the intercostal muscles, the diaphragm and the muscles of the larynx and tongue (see Fig. 8).

Farmers who permit the carcasses of their animals, especially sheep, to remain unburied and scavenged by dogs are responsible for the perpetuation of hydatid disease. This disease is the serious condition in man that results from the growth of the cyst, the stage of development that follows ingestion of the egg of a tapeworm, *Echinococcus granulosus*. The adult tapeworm lives in dogs and other canines, eggs being passed in the droppings, and accidentally eaten by a wide range of animals, including man, sheep, cattle and horses. Within these latter hosts (intermediate hosts) a cyst grows, and this cyst requires to be eaten by canines for its life cycle to be completed. In man, such cysts often reach enormous size, and if they cannot be removed by

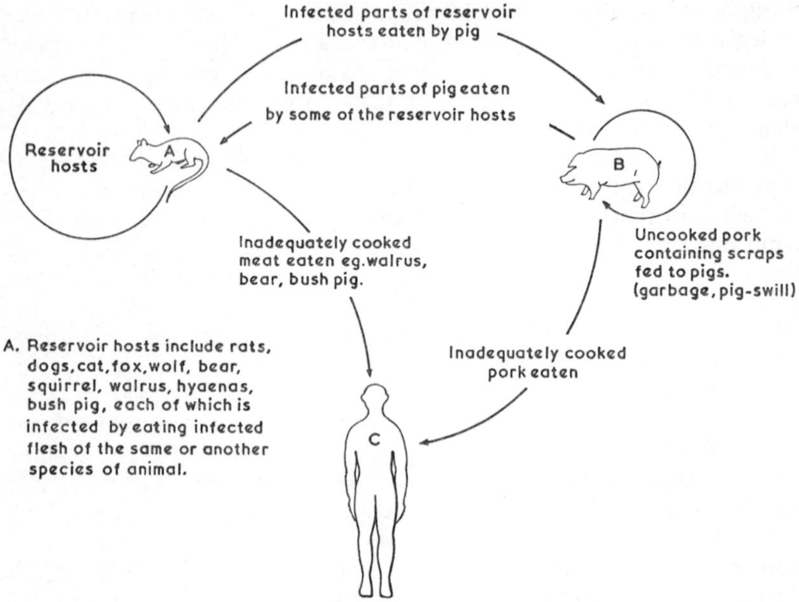

Fig. 8 Life cycle of the trichina worm, *Trichinella spiralis*. At all stages (A, B and C) infected meat is ingested, adult worms develop and lay their larvae in the intestine. These then make their way to and encyst in skeletal muscle, especially of the diaphragm, larynx, cheek and intercostal muscles.

surgery may incapacitate or even kill. The farmer has to break the dog-intermediate host link in the life cvcle; this can be achieved by the removal of carcasses from potential scavenging by dogs and the use of drug therapy for infested dogs. Despite this, the parasite still exists in remarkably large numbers and is by no means infrequently seen in dogs in Britain. Indeed, it is common practice on some farms to save sheep carcasses as food for farm dogs, despite the fact that this is against the law. Numerous examples of the importance of hygiene in the prevention of disease to both plants and animals are available.

The situation in plants is no less important, although it does not impinge so directly on human health. Where practicable, dead or diseased plants (or parts of plants) and any alternative host plants should be removed from a crop and burnt or otherwise disposed of to prevent the further spread of infection and the perpetuation of the invading organisms (see also p. 69). The use of selective herbicides can, in this context, be considered as an important hygiene measure. It is particularly important that diseased crops should not be harvested and stored with healthy crops, especially where the disease can spread under these conditions. Infested crops should be used as soon as possible after harvesting (e.g. carrots infested with carrot fly—*Psila rosae*) before further damage occurs in storage.

Seed cleaning eliminates the incidence of diseases which originate where a pathogen produces a quiescent stage of development in or on its host's seed. Earcockle of wheat (caused by eelworm *Anguina tritici*), eelworms of onions and teasle (strains of *Ditylenchus dipsaci*) are all seed borne.

Perennating parts of plants, or whole plants may be freed of eelworms by 'hot-water' treatment which involves the use of heat of sufficient intensity and duration to kill the eelworm pests but which does as little damage as possible to the host plant. Such a technique is used in the elimination of the chrysanthemum eelworms (*Aphelenchoides* sp.), strawberry eelworms (*Aphelenchoides* sp. and *Ditylenchus* sp.) and eelworms of bulbs (*Ditylenchus* sp.) (see also p. 67).

In this context it is also important to use healthy plant material for both seed or vegetative propagation. The Ministry of Agriculture in the U.K. runs certification schemes for potato 'seed' production, and also for blackcurrants and strawberries in an attempt to eliminate eelworm diseases as well as the virus diseases which the eelworms transmit.

2.2.5 Rotation

Husbandrymen involved in both plant and animal culture have long realised that a system involving the repetitive growing of one species frequently results in a loss of vigour and that the institution of a system of rotation generally offsets and may even eliminate these losses. There are many reasons why land may become 'sick', for example the depletion of certain nutrients, or the destruction of soil structure, but the accumulation of pests as a result of continuous or too frequent culture of a particular species is a cause of 'decline' in yield of both crops and animals. Rotation then is a fundamental feature of farm practice, the extent and nature of which should vary with the economic interests which in the long term are of greatest benefit to the crops and animals. The short term economic interests may, on the other hand, be most appealing and the husbandryman must take responsibility for the consequences. Vast areas have been lost to cultivation by erosion following over-grazing, or ill-advised cultivation procedures. In some instances, this responsibility cannot be allowed the farmer, and legislation has been introduced. This is seen in the sugar-beet eelworm orders of 1960 to 1962 (see also p. 25) which enforce crop rotation on all land known to be infested, and also in certain scheduled areas which may include both infested and some 'clean' land. Where the growing of susceptible crops fails to reach an economic return then a system of rotation has to be introduced and non-susceptible crops have to be grown on the infested land, thus ensuring that the disease declines below a certain level before attempting to grow susceptible crops again. Thus, where such considerations are germane it is necessary to understand those factors involved in the survival and perpetuation of pests from one year or season to another and from one crop to another.

The need for crop rotation and the care needed in devising a suitable system is seen particularly well in the case of attack by various eelworm species. Some eelworms (e.g. *Ditylenchus dipsaci*), have a wide host range, infesting potatoes, clover, onions, bulbs, strawberries as well as many other economically important plants, including weeds. Others such as the potato root eelworm (*Heterodera rostochiensis*) (see Fig. 9 and Plate 2B) and sugar-beet eelworm (*Heterodera schachtii*) (see Plate 2A) infest a limited number of solanaceous plants and species of *Beta* (and also some Brassicas) respectively. Each species, or sometimes strain, of eelworm varies in its ability to survive and multiply on particular hosts, and whole gradations of hosts exist from those on which the eelworm reproduces rapidly to those on which its multiplication is greatly reduced. Such features must be considered when formulating crop rotations. It is obviously important to remove all plants from a susceptible infested crop at harvest and prevent their remaining as 'ground-keepers' to infest future crops.

Rotations on many farms where livestock are kept involve the use of grass leys, the duration of which depends upon the needs of the particular field or farm involved. Land is sown with what is basically a grass and clover mixture, the exact composition of which is determined according to whether the mixture is required for grazing, for hay, or for silage, and also for the length of time the ley is needed. Ultimately the farmer can 'cash in' on the fertility that has accumulated during the time the ley has been grazed by animals, by ploughing and subsequently growing cash crops. If the ley is of long duration then wireworms and other grassland pests are more likely to be a problem for the crop sown after the grassland has been ploughed up, since such pests readily attack alternative plants once their usual source of food is lost.

The wheat bulb fly (*Leptohylemyia coarctata*) lays its eggs in bare soil in Britain in July, and a system of rotation should be selected which avoids the necessity of planting susceptible cereal crops, especially wheat, in such land the following year, otherwise serious losses can be expected. The inevitable presence of bare soil in July may occur during certain rotational procedures, and the growing of crops such as potatoes and beet leaves plenty of bare soil between plants.

The situation for domestic animals is broadly similar, but complications arise due to their mobility and ability to acquire immunity against a vast range of pests. Thus, providing domestic animals are not overcome by massive invasion, and the attack is not assisted by factors such as malnutrition, 'stress', or intercurrent infection (see sections 2.2.3 and 2.2.4), they are subsequently able to survive very well in the continued presence of the pest or pests (i.e., the pest or disease is endemic). If a susceptible animal (one that has no natural resistance or acquired immunity to specific pathogens) reaches an endemic area from a non-endemic area, then it may

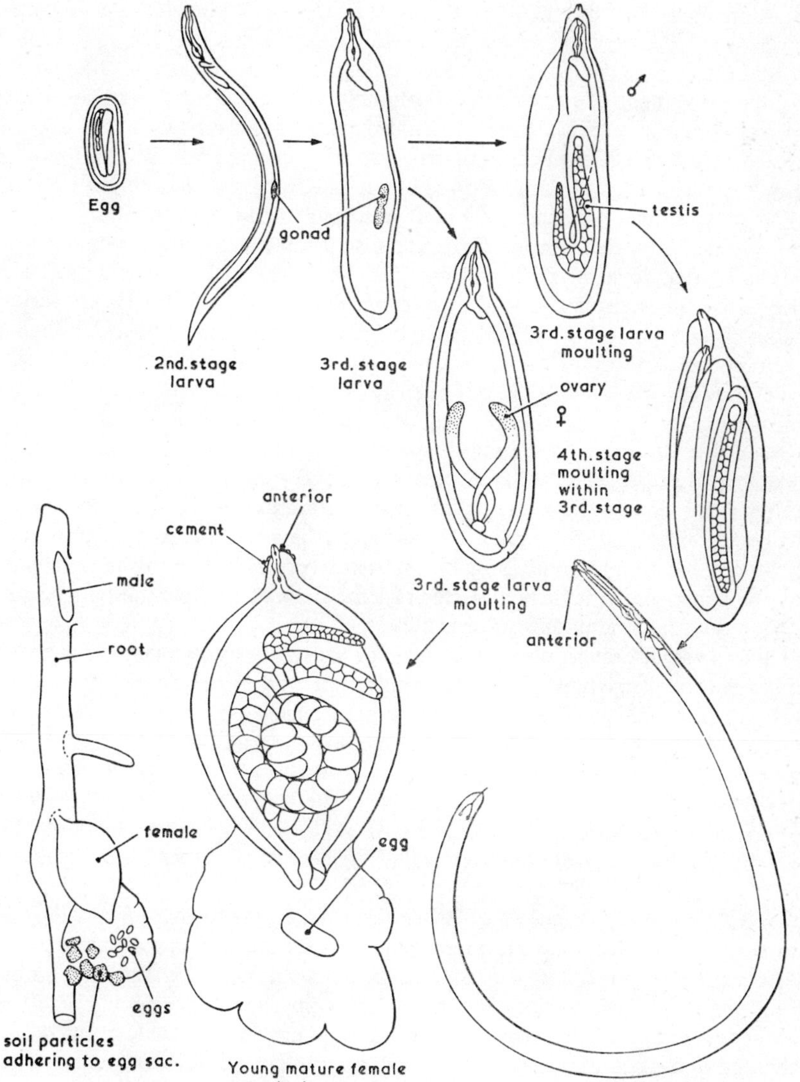

Fig. 9 The life cycle of a cyst nematode, *Heterodera* sp.

suffer very severely due to the overwhelming effect of the parasite invasion to which it is now subjected; but an animal that is continually being infected with pathogens to which it is immune will, in effect, be continuously immunised against them. All populations of man and animals have their own characteristic disease complex to which a population may be immune, or relatively so.

The movement of personnel and animals carries with it the risk of transmitting diseases to areas which previously did not suffer them, and also the clear risk of the patient becoming infected with diseases which it, in turn, has never experienced. The importance of the phenomenon of immunity is dealt with more fully in section 2.3 but it is important here to mention that the farmer has to judge for himself (in the absence in most instances of specialist advice), what level of pest invasion can be tolerated by specific groups or individual animals, so that they acquire immunity but suffer no serious reduction of vigour. Ideally, he has to prevent the uptake of a debilitating parasite burden, but at the same time permit the establishment of a 'protective' immunising dose. This can only be done, in the absence of therapeutic measures, by limiting the possible ecological association of parasites and susceptible animals. Under most agricultural conditions it is virtually impossible to segregate pests from hosts, especially such ectoparasites as ticks, mites and blood-sucking insects, and the routine use of chemicals has proved to be the only feasible method of control. Many pathogens are transferred from one location to another by birds, insect vectors, other animals, vehicles, or the wind, and it is only a matter of time before pest-free animals become infected with some pathogen or other.

There are, however, numerous diseases caused by protozoa and helminths that can be prevented from reaching economic proportions by utilising rotational procedures. Most important roundworm parasites infesting the alimentary tract of sheep, cattle and horses are passed in vast numbers as eggs in the droppings (see Fig. 10). Outside the host they undergo stages of development, each separated by a moult, the speed of which is dependent upon environmental conditions, and it is these conditions which largely determine the normal geographical distribution and hence incidence of specific parasites. All these parasites reach an 'infective' stage of development when they must be ingested or otherwise gain access to the host; if they fail to contact the host they die prematurely. After further development in the host the parasite reaches its definitive position and later the fertilised females produce eggs. The survival of such parasitic species is dependent upon their ability to withstand the period of development outside the host, from the time the egg is deposited in the faeces until the infective stage finds its way into another host. Each parasite has its own peculiar environmental requirements for development and this is one of the main reasons why the incidence of parasitism by particular species or groups of species tends to be seasonal. Modern intensive systems of management, which involve the housing of animals during all periods of the year, select those parasites which are able to survive under such conditions. Since housing tends to create fairly uniform environmental conditions for the animals which are so kept during all

Plate 1A LOCUST SWARM
(*Courtesy The Anti-Locust Research Centre*)

Plate 1B GLASSHOUSE RED SPIDER, *Tetranychus telarius*
(*Courtesy The Glasshouse Crops Research Institute*)

Plate 1C WART DAMAGE TO APPLES CAUSED THE APPLE CAPSID, *Plesiocorris rugicollis*
(*Courtesy Shell Photographic Unit*)

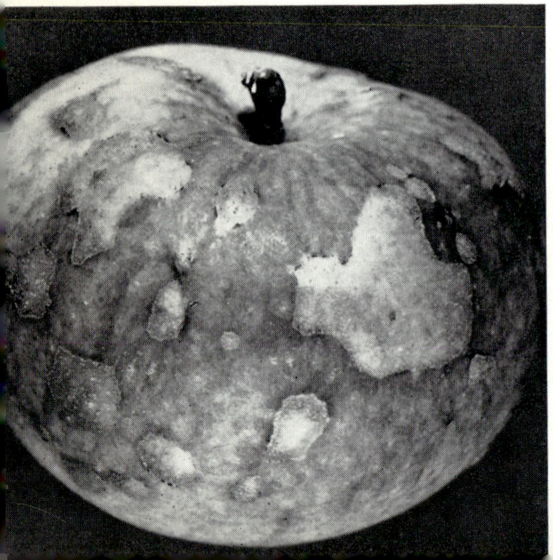

Plate 2A SUGAR BEET INFESTED WITH THE SUGAR BEET EELWORM, *Heterodera schachtii*. Note the development of 'hunger roots' which bear eelworm cysts
(*Courtesy Broom's Barn Experimental Station*)

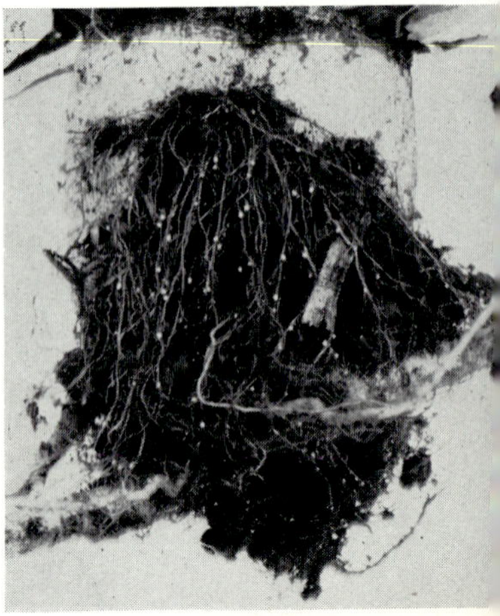

on to pasture that has not been grazed recently by heavily infested animals, and using devices such as the easily movable electric fence (see section 2.1.2) to ensure that animals do not have access to heavily contaminated pasture. In practice, however, the situation is often complicated by other limitations such as availability of land for grazing and by multiple infections with parasites which have different environmental requirements for development.

It is also complicated by the fact that in Britain and many other temperate regions of the world many species of parasitic nematodes of ruminant animals are able to survive as free-living forms in large numbers over the whole grazing season, which makes short term rotational procedures inappropriate.

Rotation is also seen in pig and poultry production where animals may be intensively housed on pasture. These animals are moved in units every few days, so that they do not infest the land too greatly with pathogens which become available for other animals to pick up once they have become 'infective'. Besides worm parasites, coccidiosis (a disease condition caused by coccidia of which the genus *Eimeria* provides most examples of importance in domestic animals) may be similarly prevented from reaching economic pest status. In short, the climatic and microclimatic factors have an important influence in the perpetuation of pests and the diseases they cause, and it is only by a study of these factors in conjunction with the biological requirements of the pests that we can formulate adequate control measures in the absence of chemical control.

2.2.6 Tillage and some other effects of management

The effects of ploughing, rotovating, harrowing and digging on pests is not always understood, despite the fact that such operations are designed to produce optimum conditions for the growth of plants that are subsequently sown. Tillage operations affect all soil inhabiting pests including (a) those which do direct damage, such as eelworms, chafer larvae, wireworms and leatherjackets, and (b) those animals such as earthworms, chafer larvae and oribatid mites which live in the soil or herbage cover, and which are intermediate hosts for specific parasites, and (c) those free-living stages of animal parasites (see p. 49) which depend for their survival upon the microclimate provided by the soil and its herbage cover.

Pests which are dependent upon such environments are killed in enormous numbers either directly as a result of the tillage operation or indirectly when they are accessible for predation by birds, or subjected to desiccating influences.

If the tillage operation removes the normal food source of the pests then they may attack the crops that are subsequently sown. Click beetle larvae (wireworms, see Fig. 12), and larvae (leatherjackets) of tipulid flies or daddy-longlegs are two pests which cause enormous damage to such crops. Many farmers attempt to exploit the stored fertility of old grassland by ploughing it up and growing cash crops such as sugar beet, potatoes or cereals. These may be spoiled by wireworms for up to three years following the initial

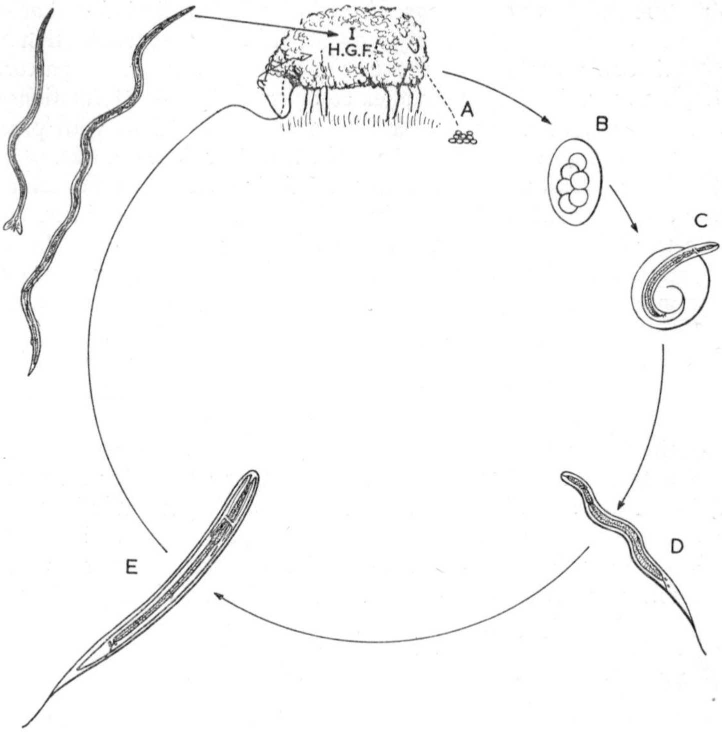

Fig. 10 Life cycle of a parasitic nematode of the family Trichostrongylidae. Members of this family infest sheep, cattle and various other animals, and include the large stomach worm of sheep *Haemonchus contortus*. The eggs are passed in the dung and subsequently a 1st larval stage emerges. This feeds on bacteria, grows and undergoes two moults before the 3rd stage develops. This is the infective stage and must be eaten before the parasite will develop further. Once within a suitable host, the 3rd larval stage emerges from within the sheath of the 2nd stage; it assumes a parasite mode of life and after three more moults becomes mature.

seasons, it follows that the parasites which survive and multiply there will show little or no seasonal incidence. Such variations as they show will be dependent upon the physiological conditions of potential hosts and the varying standards of management which influence these conditions. In general drying winds and ultra-violet rays of the sun drastically reduce the populations of the infective stages of nematode parasites and, conversely, relatively cool, moist conditions favour their survival. Since a parasite requires a combination of specific environmental factors for development it is obviously good practice to prevent animals from contacting infective stages when they are at their maximum number, and before environmental exigencies have taken their toll. This can be done by rotational grazing, moving animals to clean pasture or

PLATE 3B COLORADO BEETLE POSTER
(Courtesy The Ministry of Agriculture, Fisheries and Food)

PLATE 2B FIELD DAMAGE TO POTATOES ATTACKED BY POTATO CYST EELWORM, *Heterodera rostochiensis.*
In otherwise healthy crops of high yield potential, the potato cyst eelworm seldom causes patches which show the distinct line between 'good' and 'bad' areas such as that shown in the photograph. Cyst eelworm infestations are frequently complicated by other factors, such as *Rhizoctonia* (a fungus) infection and it is under such circumstances that the distinct crop patchiness occurs. (*Courtesy A. L. Winfield, M.A.F.F., National Agricultural Advisory Service, Eastern Region, Cambridge*)

PLATE 3A THE EFFECT OF A 'TRAIL' PHEROMONE ON TERMITES (*Courtesy C. B. Moore*)

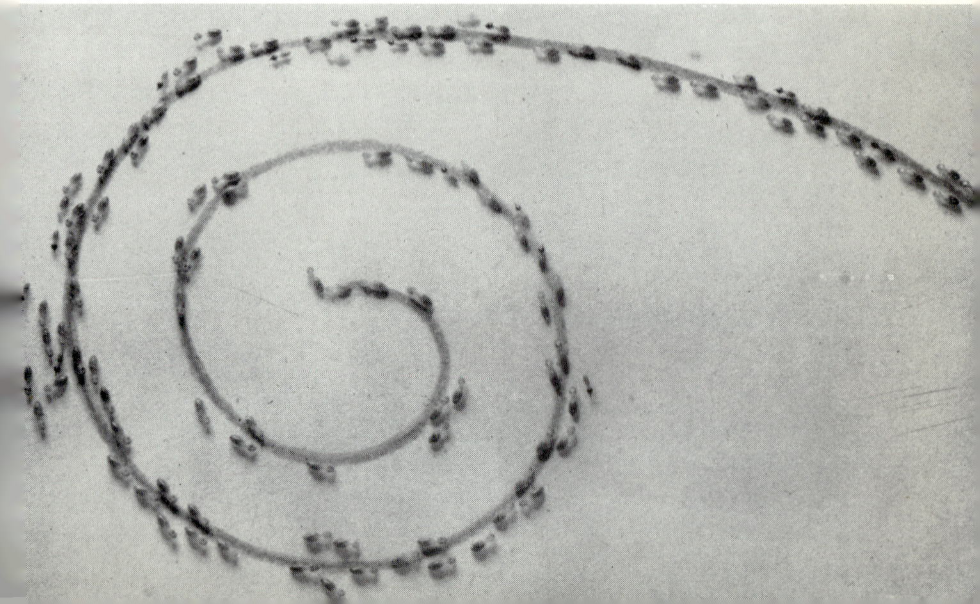

ploughing. This is the time taken for the various generations of wireworm larvae to become adult and leave the soil. Leatherjackets may also attack sugar beet and other crops but under normal conditions they cause most damage to spring sown cereals, especially oats, and newly sown grasses.

Foreknowledge of pest attack by such organisms enables tillage operations such as ploughing to be carried out at times of the year when the greatest pest destruction can be achieved. Autumn ploughing is recommended to control leatherjacket infestation as the female fly lays her eggs in grassland in autumn and ploughing at this time will remove potential sites for oviposition. On the other hand it is alleged that the wireworm population is equally reduced irrespective of the time of ploughing.

Whenever possible, the effects of particular tillage operations upon particular pests and the time(s) of year when these are practised should be known, and should single or combinations of operations significantly and economically reduce the pest population, then they should be practised. For instance, it has been shown that intensive and deep cultivations readily reduce the population of migratory eelworms (*Trichodorus* sp.) and that following such operations the damage to subsequently grown beet is considerably lessened.

Ploughing and rotovating also affects the intermediate hosts of certain parasites of farm animals by direct or indirect population reduction. For instance, the population of oribatid mites, which are intermediate hosts of the sheep and cattle tapeworms, *Moniezia* spp., and of the chafer grub, intermediate hosts of the thorny headed worm of pigs (*Macracanthorhynchus hirudinaceus*), are reduced by cultivation. When herbage is buried during ploughing many free-living stages of parasites which infest animals are also buried, and unless they migrate back through the soil on to newly established herbage cover they fail to complete their life cycle. One may well recommend, if economically feasible and desirable, that a pasture that contains a heavy infestation of the eggs and free living larval stages of parasites should be ploughed and a crop grown to which animals shall not have access. Alternatively infested pasture may be utilised for hay or silage, under which conditions the parasites will die.

Environmental factors influence the rate of development and longevity of the free-living stages of many gut dwelling nematode parasites of grazing animals (see also p. 48). The faecal pat harbours all stages of development and buffers them against adverse weather conditions, particularly the drying action of sun and wind. So long as the faecal pat remains the infective (3rd stage) larvae which develop there are able to migrate onto herbage when weather conditions are appropriate; the earlier larval forms, which are less resistant to adverse weather conditions, remaining. Chain harrowing, to spread faecal pats, if done in drying weather, will cause a heavy toll of all stages of larval development but especially of the 1st and 2nd stage larvae. If harrowing is done in wet, cool weather then parasites will not only survive but be most effectively disseminated.

2.2.7 *Planting techniques*

(a) VARIATION IN PLANTING TIMES

The time at which one plants a crop is determined by the cultural requirements of that crop so that maximum economic advantage may be achieved. Numerous factors influence the time of planting, but where crop damage is anticipated it may be possible, by varying the date of planting, to ensure that the crop is either at its least susceptible stage of development when attack is likely to occur, or that it is saleable before the pest can exert its effect.

Pest attack synchronises with the availability of a potential host plant, but breeding techniques have produced varieties of crops that extend the normal season of growth. Increased knowledge of crop and pest biology has enabled us to know which crops are most susceptible, which stages of growth are most susceptible, and which crop and climatic factors influence pest development. In short, increasing knowledge permits increased awareness of the factors involved in the epidemiology of infection.

An excellent example of the use of this technique has been recorded from the early potato growing area of Ayrshire where the potato root eelworm is also found. In this relatively mild part of Britain it has been found possible to produce an economic yield from the variety Epicure by planting in mid-February and harvesting not later than the 20th June. Epicure grows significantly at temperatures lower than 7°C (which is too low for main crop potatoes such as Majestic and King Edward's) when the eelworm is inactive. By June when the crop is harvested the eelworms have invaded the crop but are unable to complete development beyond the immature 'white cyst' stage. These do not contribute significantly to subsequent population levels because very few of them contain viable eggs. In Belgium the variety Eersteling is used similarly.

The carrot fly (*Psila rosae*) normally produces two generations each year in Britain; the first egg laying adults appear from mid-May to June and the second from the end of July to September. The first generation flies usually emerge from fields sown with carrots in the previous year, whereas the second generation flies develop from eggs laid by the first generation flies in fields currently sown with carrots. Where carrot fly damage is expected therefore, carrots can be sown in the latter half of May or early June so that germination is delayed until after the first generation flies have ceased to look for hosts. Crops then suffer comparatively little damage from second generation flies unless nearby fields have been infested earlier.

(b) SPACING OF PLANTS AND CROPS

The optimal planting distance between plants is that distance which permits the crop as a whole to produce the greatest financial return for the investment of labour and materials. The spacing of plants in a crop depends upon the requirements of that crop and the ability of the soil and climate to meet these demands. A change in any one of the wide range of contributory factors that operate in this system may be compensated or exaggerated according to the limitations of the system. Thus, if a crop is attacked and damaged at its roots (e.g. brassicas by the cabbage root-fly *Erioischia brassicae*) then in a rich, moist soil the plant may not only survive but also produce an economic yield. In a poor dry soil the plant may irrevocably wilt, and die. If the crop population is so high that the rate of evaporation and transpiration exceed the precipitation as rain (or soil water through irrigation) then such a crop is more likely to wilt as a result of root damage than one where the plants are so dispersed that they have adequate nutrients and water.

A knowledge of pest behaviour, and especially its habitat preferences, its ability to locate a crop or animal and the distance it will travel to do so, can be made use of in formulating preventive control measures. The distance between susceptible crops may be important where readily distributed insect pests are involved and some, such as aphids, may be readily transported by currents of air over vast distances. Other insects, such as the carrot fly, usually travel short distances only and this particular pest lays her eggs

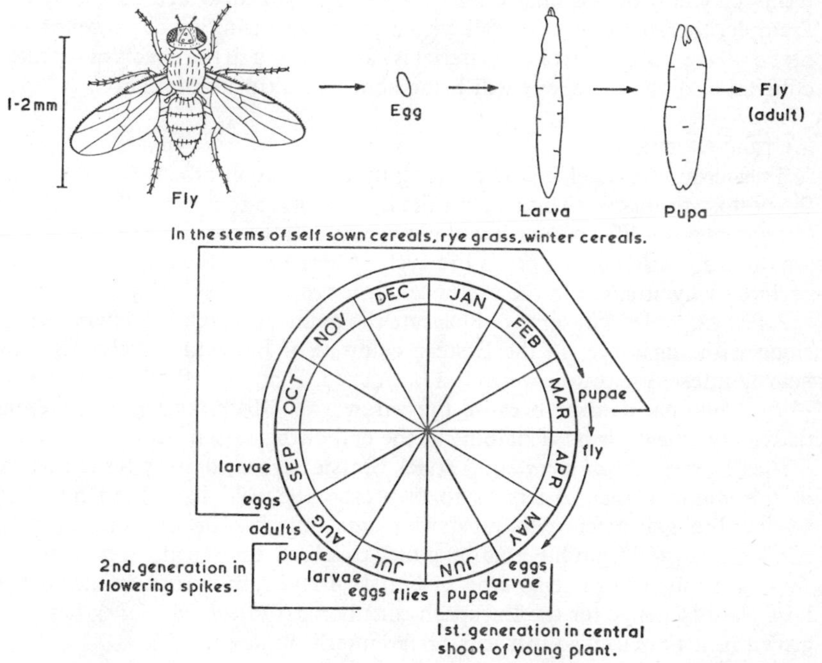

Fig. 11 Life cycle of the frit fly, *Oscinella frit*, a common pest of cereals and an inhabitant of grassland. In the United Kingdom there are typically two summer generations each year.

mainly around the plants at the edges of a field since it spends most of its adult life sheltering in hedgerows. Under such circumstances it is obviously sound policy to grow large fields of carrots rather than a number of small ones. Similarly, the frit fly (*Oscinella frit*) (see Fig. 11) though not venturing far into an open field may travel considerable distances from one field to another in the shelter of hedgerows, and it is also well known that hedgerows may induce wind eddies behind them in which flying aphids particularly are 'thrown out' onto the crop.

In regions where the pseudococcid, *Pseudococcus njalensis*, transmits 'swollen shoot', a virus disease of cocoa plantations, it has proved valuable to space the trees at such a distance that their branches do not interlace. The coccids, normally immobile, are carried from tree to tree by ants which tend the coccids, feeding on their sugar rich excreta. Interlacing branches of adjoining trees are a ready route for dispersal.

A high seeding or planting rate may also be useful where relatively immobile insects such as those infesting the soil are involved. High seed planting rates are recommended where damage to crops by wireworms (*Agriotes* spp.) (see Fig. 12) is expected. In effect this lessens the amount of damage, in terms of yield, of the final crop but does nothing to affect numbers; this example, therefore, does not fall within the strict definition of prevention of pest damage since more plant material is eaten, although the effects of repeated cultivation discussed above will bring about an ultimate decline.

(c) TRAP CROPPING

Trap cropping involves the planting of a crop upon infested land so that the pest is stimulated to attack, but the crop is either removed before the pest can complete its life cycle or it will not provide all the requirements necessary for the completion of the pest's life cycle. Alternatively the trap crop may be preferentially attacked in the presence of the crop one needs to protect.

A few examples will serve to illustrate the point. A method of sowing wheat along with sugar beet in the Eastern counties of England in fields that are heavily infested with wireworm has been used. It appears that the sugar beet suffers minimal damage because the wireworms attack the wheat preferentially, any wheat plants remaining at the end of the season being removed.

Flea beetles (*Phyllotreta* spp.) attack brassica crops, destroying the leaves of the young plants. Attempts to trap crop the field have been made by sowing the perimeter with mustard plants (another brassica) so that the beetles emerging from hibernation attack this first. Insecticidal treatment can then be applied to the trap crop *in situ*. In France, mugwort, an alternative host plant to maize for the European corn borer (*Ostrinia nubilalis*), has been grown near to maize, so that it is preferentially attacked.

In Holland, *Tagetes* spp. have been used as a trap crop following a potato crop that has been infested with potato root eelworm. In this instance, the trap crop induces potato root eelworm eggs to hatch but since it will not

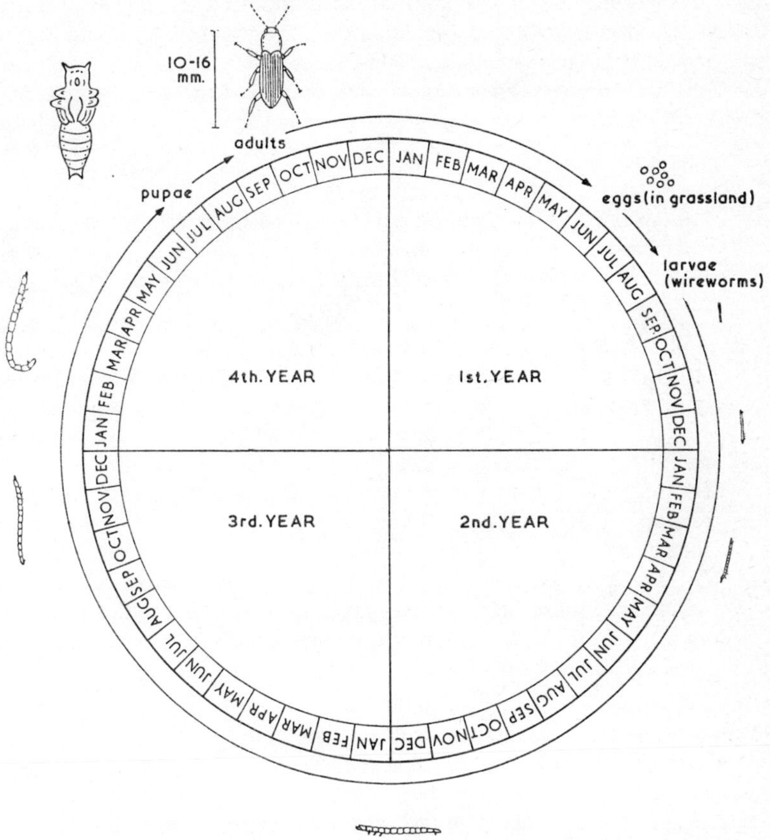

Fig. 12 Life cycle of click beetles, *Agriotes* spp. Damage typically follows the ploughing up of grassland when the polyphagous wireworm (larva) turns to arable crops for food. Because of their long life span, injury can persist for 3 years, the amount of damage increasing as the larvae grow larger.

support larval development, the young forms die. The plants are then ploughed in as green manure.

2.3 Resistant and Immune Hosts

Resistance is a term used later (see p. 165) to describe the ability of pests to survive and multiply in the presence of pesticides, but in this section its use is related to the innate and/or acquired resistance of the host to infestation and to the debilitating effects of infestation.

Resistance and immunity are sometimes used synonymously to encompass all those intrinsic factors that enable a plant or animal to withstand or overcome infections or infestations, and sometimes to differentiate between innate (=natural = non-specific) resistance and what is known as acquired (or specific) immunity.

(a) INNATE RESISTANCE (Natural resistance)
Due to host physiological and anatomical factors which are genetically determined. Factors which affect the physiological and anatomical features affect the degree of innate resistance.

(b) ACQUIRED IMMUNITY (refers to animals only)
(i) *Active.* The host forms immune bodies as a result of exposure to antigens produced by an invading organism or other foreign immunogenic material. The immunity may persist throughout life, or wane in the absence of further stimulation, and is either naturally acquired by sublethal exposure to the antigen(s) or artificially induced by injection of immunologically active antigen(s) which promotes the formation of homologous antibody.
(ii) *Passive.* The host does not participate in the formation of immune antibodies, these being either naturally acquired via the placenta or colostrum or artificially induced by injection of immune bodies produced by another host.

Basically, natural resistance (called here simply resistance) is due to natural anatomical and/or physiological features of a plant or animal that are under genetic control and which prevent a particular pest from invading and/or completing some or all of its obligatory life processes. These features may be modified by a variety of conditions to which the animal or plant is subjected, including diet, age, intercurrent infection, sex and 'stresses' of various kinds (see also p. 38). Therefore, within a population of animals or plants there is great variation between individuals in their resistance to particular pathogens.

Living things are continually exposed to a wide range of pest species, but only a fraction of these species is able to invade and undergo development in any particular host. This means that each species (or even strain) of plant or animal has its own particular pests which may be quite distinct from those of other host species. In other words, each pest may have either a narrow or a wide range of potential hosts (a narrow host specificity or a wide one), and the host may provide some or all the stimuli and conditions required by the pest for its existence.

Animals and plants within particular geographical regions may have pests which are characteristic of those regions. This does not necessarily mean that these animals and plants are not susceptible to pests found on or in hosts of the same species elsewhere, but they may not come into contact with the pest. Thus cattle in Europe, for example, are highly susceptible to African trypanosomiasis but do not contract it because the tsetse fly vectors (*Glossina* spp.) and other biting fly vectors are unable to survive outside a limited geographical range.

Innate resistance is a phenomenon found in both animals and plants, and in particular areas use is made of those organisms that are resistant to pests that are endemic. It is obvious folly to attempt to keep susceptible crops or animals in regions where they are liable to be seriously debilitated or die as a result of infestation, and the ability to keep them often depends either upon developing resistant strains or on continuous pesticide applications.

Brief mention is made later (p. 65) of the breeding of animals and plants for resistance to particular pathogens, and in this way man is attempting to accelerate the evolution of genetically controlled defences. Resistance to many noxious agents may be conferred by mutations of single genes which have occurred naturally during the evolutionary process. The beneficial character afforded by such a mutation is reinforced in a natural population because of the advantage it confers to those individuals that inherit the character. Clearly the colossal range of genetic variability, expressed in different species and races of animals, and varieties of plants, testifies to the diversity of evolutionary pressures which caused their production, and this leaves man with an enormous genetic reserve which he may use in producing new strains of animals and plants. Besides this natural reserve, techniques are now available for inducing mutations by chemical and radio-active means, though this has found greater application in the breeding of plants. It should, however, be remembered that pests too are capable of similar adaptability and this leads to very serious problems and complications as we shall see when discussing the development of resistance to pesticides (p. 165).

Interesting cases of resistance to *Plasmodium falciparum*, the causative organism of malignant tertian malaria accompany both the so-called 'sickle-cell' gene which is common in some African, Asian and southern European populations, and also the sex-linked inherited abnormality of the red blood cells of many negroes which suffer a deficiency of the enzyme glucose-6-phosphate dehydrogenase. The geographical distribution of these two red blood cell abnormalities corresponds with the geographical distribution of 'falciparum malaria'.

The sickle-cell condition is manifested in the production of abnormal haemoglobin (the respiratory pigment of red blood cells), so that the oxygen-carrying efficiency is reduced and the cell is distorted into a characteristic sickle-shape. In homozygous individuals nearly all the haemoglobin is of this abnormal type whereas in the heterozygotes much less is in this form. The homozygous condition therefore imparts lethal characters to the red blood cells so that individuals rarely survive to adulthood. Despite this there is a high frequency of the sickle-cell gene in populations living in the above areas of the world and it appears that this trait confers protection against 'falciparum' malaria. The exact reason why the sickle-cell is not a suitable medium for the growth and multiplication of 'falciparum malaria' is not known but it appears likely that the parasite cannot metabolise the abnormal pigment satisfactorily.

The glucose-6-phosphate dehydrogenase deficiency (or relative deficiency) of red blood cells is very disadvantageous under natural conditions and finds its full expression in males who have abnormal X and normal Y chromosomes (hemizygous males) and partial expression in heterozygous females (having one normal and abnormal X chromosome). The degree of natural resistance to the malarial organism appears to correspond with the relative deficiency of this enzyme, and again it is not known why this should be so.

Differences in the resistance of plant varieties to insect attack have been known for many years and have been widely used as a means of controlling certain infestations. A classical example of this is afforded by the infestation of grape vines by the grape aphid (*Phylloxera vitifoliae*) which very nearly ruined the French wine industry during the latter part of the nineteenth century. This economic and social catastrophe was only averted when *Phylloxera*-resistant root stocks were sent from the U.S.A. to France. Cotton can be grown in Africa because a strain has been bred from a single plant that was resistant to leaf-hoppers of the genus *Empoasca*.

Painter (1951) attempted to classify the various mechanisms of plant resistance under the following headings. Though these are inter-related, each mechanism may exert an effect on its own or in company with either or both of the other two.

(i) *Preference* (or non-preference) is concerned with plant characters and insect responses that cause a pest to seek out or ignore a particular plant for oviposition, for food, shelter, or for combinations of all three. The colour, smell, taste, thickness, toughness, hairyness or waxy nature of a plant may either attract, or repel, or fail to attract a particular species of insect.

(ii) The term *antibiosis* refers to those plant factors which adversely affect the life history of a pest when it uses a particular variety or species for food. Even if the pest survives, the resistance of the plant may be indicated by reduction in pest fecundity, decreased pest size, increased pest mortality or abnormal length of life. It is still not clear how antibiosis is manifested in most instances, though it appears that the presence of specific chemicals or lack of particular nutrients in the host plant regulates the life of the pest to some extent and hence confers some degree of resistance to the plant. One example of such a plant regulatory mechanism has recently been discovered, this being that certain plants which are resistant to particular insects produce highly specific chemicals which are related to, and exert a similar effect to that of the juvenile hormone which the insect produces. Such a substance has been described from the tissues of the American Balsam fir.

Juvenile hormone, produced by two glands (corpora allata) in the head, is an absolute requirement for immature insects if they are to progress through the usual larval stages. Before the larva metamorphoses into the adult, the flow of this hormone must cease. The juvenile hormone may be secreted later in the adult of some species, but only after they are mature and it must be absent from insect eggs if they are to undergo normal embryonic development. Plants producing substances similar to juvenile hormone will therefore prevent development of insects for which the hormone exerts a specific effect. It appears likely that several plants may possess substances related to juvenile hormone which confer on them resistance to particular pests. Isolation and synthesis of these substances may have considerable potential as pesticides, especially since their activity is specific to insects and appears to exert no effect upon other forms of life.

(iii) *Tolerance* differs from preference and antibiosis in the predominant part played by the plant. Here we are concerned with all the factors involved in the ability of plants to repair or replace parts and thus minimise the effects of pest damage, and to consider how these factors are influenced by the environmental conditions in which the plant grows. The more vigorous the growth of the plant, the more likely it is to be able to replace damaged or lost tissues. Tolerance may be associated with the ability of the plant to readily produce callus to heal wounds, to neutralise toxins produced by invading organisms, such as eelworms and aphids, or to produce galls or liquified tissue to wall off and isolate the invading organism. Many other examples exist and are reviewed by Painter (1951).

A recent review of some of the factors that influence the plant-insect relationships has shown how a knowledge of these aids an understanding of the phenomenon of resistance in plants. For even a low level of plant resistance may bring about effective pest control when integrated with the action of natural enemies (see p. 154). A good example illustrating some of the various factors that influence resistance is that of the frit fly (*Oscinella frit*). A very young plant presents little obstacle to entry by the newly hatched frit fly larva but such a plant may provide insufficient nutrient for the pest. If, on the other hand, the plant is sufficiently well grown and can provide adequate food for the insect's development its tiller surface may be too hard to permit larval entry. The best time for larval invasion is therefore between these two growth phases, so that if the presence of a uniform susceptible stand of oats, such as occurs when a crop is sown in spring, coincides with the maximum incidence of insects, excellent conditions are provided for the pest. The frit fly also occurs on a range of grasses, but the variability in stages of grass growth under natural conditions means that comparatively few tillers will, at any one time, be at a suitable stage of development for invasion. These factors are not the only ones influencing resistance to this pest, since certain varieties of oats are not attacked by virtue of the 'non-preference' shown by the ovipositing female fly towards varieties possessing an unusual colour or a particularly hairy structure.

The phenomenon of *acquired* or *specific immunity* is exclusive, so far as we are aware, to the association of animal hosts and their pests, though how widespread or significant the phenomenon is in animals other than vertebrates is not well documented. Here we are concerned solely with acquired immunity in mammals and birds. It depends upon the ability of the host to respond to the presence of the pest by producing complicated but specific chemicals (antibodies). Thus when an organism is invaded specific cells react to the invasion and a state of 'specific altered reactivity' is established. The consequences of this altered reactivity depend upon circumstances and may be beneficial, as in the production of immunity, or harmful, resulting in disease or clinical hypersensitivity. It is most important to stress this point since certain parasite infections may induce either state. For example, young bovines initially infected with small numbers of the lungworm *Dictyocaulus viviparus* suffer a mild pneumonia from which they can be expected to recover, unless secondary bacterial infection ensues. This acquired immunity can be expected to be beneficially boosted when the animal continues to pick up a further small infection from pasture. On the other hand the ingestion of large numbers of parasites in previously lightly infected animals may in certain circumstances provoke clinical hypersensitivity and death. The quantitive increase in reactivity of hypersensitive cells results in lung congestion which effectively inhibits adequate ventilation. Similar hypersensitivity reactions to a number of nematode infections are known though none are so dramatic as that described. The consequences of any infection depends, therefore, upon local circumstances and the outcome of any infection is largely governed by the interplay of host reactivity and parasite activity. Clearly the site of interaction is of paramount importance, the relative sensitivities of different organ sites varying for different animals.

In the establishment of the beneficial immune state the pest as well as its metabolic products are neutralised by cell-produced antibodies, while other cells remove the invading organism, or its breakdown products, by phagocytosis. Still other cells will be involved in healing and replacing injured tissue. Acquired immunity may, depending upon the invasive organism involved, and the intensity of infection, be short lived or may persist throughout life. It may also be local or generalised throughout the body, and may be a 'sterile immunity' whereby not only are the parasites destroyed but certain cells and their descendents remain sensitised for months, years, or even life and effectively prevent reinfection. Alternatively a state of 'premunity' may exist (non-sterile immunity) which is dependent upon the persistence of a latent infection whereby, if the pest organism disappears from the body then the immunity wanes and is soon lost and the animal may be re-infected.

Many plants have been shown to produce inhibitory substances (phytoalexins) in response to infections with fungi which then render the plant immune to further infection. Though the property is acquired in the sense that it depends upon exposure to an infective agent, unlike the condition of acquired immunity described in animals, the phytoalexins are not entirely specific and inhibit a wide variety of potential fungal attacks. The importance of phytoalexins to animal pests is not known.

When protozoa, helminths, or arthropods infest an animal certain basic differences in the pattern of infestation become obvious and these determine, to a large extent, the nature of the immune reactions. Any pest which comes into intimate contact with a mammal or bird presents a series of complicated chemical products (antigens) against which the host reacts by producing antibody. In other words the host is able to recognise the difference between what is self and what is foreign. The rapidly multiplying protozoa represent an enormous antigenic increase and are able to multiply until the host has produced sufficient antibody to inhibit their reproduction and remove the parasites from its system. Helminths, on the other hand, do not generally multiply in the body of the definitive host and they go through a series of developmental stages often accompanied by specific migrations through the body before attaining sexual maturity and taking up their definitive positions. They are more complex structures than protozoa and produce a more complex and sometimes different series of antigens at different stages of development, all of which incite the production of host antibody. Since, however, the vast majority do not multiply within one host the amount of a particular antibody finally produced against any one stage of development will in general be small. Further complications arise which are basic to an understanding of parasite invasion. Firstly the host reacts to the 'mosaic' of antigens by producing a mosaic of antibodies, but only certain of these are protective or functional in action, the majority being of no protective value to the host whatever (i.e. non-functional). Secondly, the evidence suggests that it is only when the amount of protective antibody reaches a certain critical level (titre) in the body that the host shows a significant protective immunity. Thirdly, this titre is related to the degree of host tissue penetration achieved by the parasite, the less the penetration the less the degree of host reaction. Fourthly, it is commonly found that the degree of acquired immunity to helminth and protozoan infections rapidly wanes in the absence of further antigenic stimulation. An animal which has acquired a degree of immunity is therefore likely to become more susceptible to reinfection the longer it is kept free of parasites. Most protozoa of importance to man and his animals invade tissues to some extent and hence induce a considerable host reaction whereas many parasitic helminths and arthropods may spend some or all stages of their life living on the skin, or within the lumina of various organs. Thus those stages of parasite development or those parasites which invade tissue most severely can be expected to be those that induce the greatest degree of acquired immunity. For example, trypanosomes, malarial parasites and piroplasms, schistosomes (blood flukes) and adult filarial worms will in general induce a vigorous immune response because they invade the blood and fixed tissues. Parasites such as adult tapeworms, and many intestinal nematode parasites which live in the gut lumen, or insects, ticks, and mites which live in or on the skin excite little response. Parasites which inhabit both the tissues and the intestinal lumen, depending upon their stages of development, evoke a response which is related to the extent of tissue invasion. Examples of this latter type are many, e.g., hookworms, ascarid worms.

Like all generalisations however, there are a number of exceptions to these examples. For instance, the juvenile forms of filarial worms, (microfilariae) which are present in blood, appear to evoke little if any immunity.

Antiparasitic antibodies are in manv cases demonstrable in normal serological reactions and use is made of this fact in medical and veterinary diagnostic laboratories. Parasitism is a normal finding in nature and the degree of immunity is to a large extent determined by the parasite burden experienced by an animal during its life; in an immunologically competent animal the antibody titre, be it a functional or non-functional antibody, will in general increase with exposure to infection. That is assuming, of course, that the animal is not killed! It is therefore desirable where a susceptible animal is likely to be infected that it be allowed to pick up a small, non-debilitating infection at regular intervals so that it becomes fully immune. The attainment of this desirable condition is a more difficult matter and where the system breaks down drug therapy is needed.

Such are the dangers of moving susceptible animals into areas where particular diseases are endemic that it is common practice in some parts of the world to infect those animals with the more debilitating pathogens before moving them into such an area. For example, where the piroplasm *Babesia bigemina* (causing red water fever in cattle) is endemic, it is common practice to inoculate blood from infected cattle which contains the pathogen into susceptible healthy cattle. These animals then suffer the disease but are kept under observation until they recover to ensure that they are not too seriously debilitated. They can then be moved into areas where the disease is endemic and are protected against serious infection, despite the fact that infected tick vectors can continually inoculate infective organisms into them during its blood sucking activities.

Husbandrymen the world over are always keenly aware of the fact that locally produced animals are more likely to flourish better on their farms, at least initially, than animals brought in from elsewhere. The main reason for this is undoubtedly the fact that home produced animals become immune to the local combination of parasitic flora and fauna at an early stage in their lives.

The host reactions to the majority of parasitic arthropods, such as mites, fleas, ticks, biting flies, lice, will be familiar to most people. The immunological reactions are generally of a low order and animals do not in most instances appear to develop any significant degree of protective immunity.

So far, the acquired immunity under discussion has been naturally acquired immunity as a result of actual parasite infection. It is possible to induce immunity artificially by the use of vaccines derived from the killed or attenuated parasites and their metabolic products. Unfortunately, because it is not yet technically possible, protozoa and helminths have not lent themselves to the methods of vaccine production which are normal with bacteria, since the killed or attenuated organisms (accomplished by means of heat, chemicals or radiation) do not provoke an adequate protective antibody production from the host. It appears that the living parasite and its metabolic antigens are necessary for this and the technical and biological problems of attenuating the organism or defining the nature of its metabolic antigens are at the moment unsolved. Recently, however, a vaccine has been produced against the bovine lungworm (*Dictyocaulus viviparus*), by X-irradiating the infective larvae before giving them to susceptible cattle. These irradiated larvae undergo the normal complicated migration through the body and the accompanying moults but cannot produce sexually mature adults. The migration of sufficiently large numbers of larvae permits the acquisition of a protective immunity to further natural infection. This vaccine, Dictol*, is marketed commercially—its great disadvantage being that it has a relatively short shelf life since the living larvae die if not used within a relatively short period of time. A vaccine composed of irradiated *Dictyocaulus filaria* (the related species which infects sheep) has been successfully used in Yugoslavia for several years though the vaccine appears to be of little use in producing protection in lambs of under six months of age.

Antibodies may be conveyed from an immune mother to her young either through the placenta or, more usually in the case of farm stock, via the colostrum or 'first milk' that the newborn suckles. This colostrum, rich in globulin proteins aids the newborn to withstand certain specific diseases it might encounter early in its life before it has become competent, immunologically speaking, to produce its own antibody in sufficient quantity or quality.

Artificial passive immunity, the passing of immune serum from a donor to a susceptible animal, is of no significance in the treatment of protozoan and helminth infections, though immune sera are marketed for use in attempts to control ticks on individual dogs in tick infested areas of Australia.

It has been mentioned above that the newly born animal may be in a condition of what is known as 'immunological incompetence', that is the body cells responsible for distinguishing between what is self and what is foreign to the body are not fully developed and functional. This period of incompetence is usually very transitory but during this period and in the absence of maternal antibody in the colostrum, the intake of a number of parasites may not be recognised as foreign by the body, thus permitting the parasites to develop fully and provoking no (or insufficient) immunological reaction from the host (see p. 40). If the body is thus taught to accept the parasite as 'self' then its subsequent ability to resist infection is materially inhibited.

* Allan and Hanbury Ltd., Ware, England.

Thus Urquhart (1958)* found that newly born cattle in areas where the beef tapeworm is endemic (much of Africa) may be infected shortly after birth, and this may be a reason why the parasite is present in such large numbers in animals of all ages. In the immunologically competent animal tapeworm cysts are destroyed within a period of nine months and cattle are then immune to subsequent infections.

It has also been found that sheep, prior to the age of 4–6 months, are not able to develop a significant protective immunity to many nematode parasites which inhabit the gut. Heavy infections received before this period are therefore serious and have to be removed by drugs, too heavy an infection even if not mortal may swamp the immunological processes within the lambs and materially affect their ability to react competently to further infection later in life.

Each community of plants or animals throughout the world, has its own characteristic pests and diseases. Natural selection has produced plants and animals that are capable of existing in the presence of such diseases. Though the genetic constitution of such plants and animals permits their survival in regions where particular pests and diseases are endemic or epidemic it is unfortunately often the case that disease resistance does not always go hand in hand with other desirable attributes. The importation of susceptible animals or plants into an area where disease is endemic, be it on a single farm, or a larger geographical region, inevitably leads to at least a temporary decrease in vigour and often total loss through deaths. In many regions of the world, particularly in the temperate zones, husbandrymen have selectively bred for increased productivity of animals, early maturing, heavy milk yields, increased egg output, etc. These regions are also free from certain highly debilitating diseases such as trypanosomiasis (protozoa) and rinderpest (a virus), infecting domestic animals; and malaria, trypanosmiasis (protozoa), and schistosomiasis (blood fluke) infecting man. It is possibly true to say that one of the reasons for man's advancement in the temperate regions of the world is that he and his animals have not been plagued with the diseases that rage in the tropical regions. With the advancement of knowledge and of control techniques these regions of the world are advancing beyond the position of subsistence living and require a greater productivity from man, his livestock, and his land.

The majority of highly productive domestic animal breeds owe their origin to Europe and it is not unnatural that attempts have been made to introduce these animals into other regions of the world, sometimes with success but often without, especially in tropical Africa. It is now possible to vaccinate against some of the most severe bacterial and virus disease of domestic animals in areas where these diseases are endemic but it is not possible as yet to do this against the various diseases caused by protozoa,

* Urquhart (1958) cited in Soulsby, E. J. L. 1962. Antigen–Antibody reactions in Helminth infections. *Advances in Immunology* **2**, 265–308.

helminths and arthropods. Instead, drug or chemical treatment, which is expensive in manpower and which is not often feasible where nomadic and other 'independent' peoples are concerned, is the only answer at present. Attempts are also being made by breeding experiments to make the indigenous animals more productive and the European animals more resistant to these pathogens, but this is a very long term programme and such a scheme is fraught with the danger that the parasite may adapt itself to a new set of host conditions as quickly as resistance is being developed by the host. The provision of an altered environment for the parasite, especially in the case of the rapidly multiplying protozoa, whether it be due to natural antibody, drug therapy, or the advent of a resistant host exerts its selective pressure upon those variants of the pathogen that can survive and multiply in such conditions. The fact that many pathogens are able to accomplish this is one of the major hazards in any control or eradication venture (see section 8.2.1). One other difficulty is that the breeder is frequently confronted with the problem of breeding resistance to two or more pathogens.

The problems of breeding crops for resistance to various animal pests are perhaps less acute than they are for animals since the rate of propagation of resistant varieties is very much greater, and results can be expected to materialise more quickly. On the other hand, the variations within single species of pests in different localities mean that a crop may be resistant to one pest strain but not another. This is a difficulty when breeding for resistance to the potato root eelworm (*Heterodera rostochiensis*) and many other eelworm species, but despite this, good progress has been made and resistant varieties of potatoes are now being used commercially.

Many other agronomically important plants show moderate to good resistance to a variety of insects and are available for use in areas where losses are anticipated. As in the case of studies on resistance of animals to pests, so the factors that cause the pests to alight, reproduce, and feed on plants require to be studied.

Chapter Three
CONTROL OF PEST DAMAGE BY DIRECT POPULATION REDUCTION
The use of Physical Agents, Inorganic Chemicals and Plant Derivatives

3.1 Direct Reduction by Physical Agents

The use of physical agents in an attempt to control pest animals is often comparatively costly and this is one of the main reasons why chemical agents have achieved their importance. Prior to the introduction of chemicals and the technology associated with their usage, physical methods were the only means available for the control of a wide number of pests. Indeed, in vast tracts of the world, in the absence of the capital resource to purchase and train personnel in the use of chemicals and where labour is cheap, physical methods are still widely used and therefore often the sole means of control.

Many pests are still best destroyed by such time honoured physical methods as hunting, shooting, trapping, beating, etc. though advanced technological achievement has permitted a high degree of sophistication in certain directions, notably in the use of thermal and irradiation techniques. All these methods rely upon a knowledge of the biological requirements or habits of the pest in question. Shooting and trapping are widely used methods of reducing bird and mammal populations which damage man's property directly, his crops or which are predatory on his animals, or which serve as alternative hosts to a wide variety of disease-causing organisms.

Trapping is useful where pests are established in particular environments or where their pattern of behaviour lends itself to the reliable placing of traps. Many pests are caught in this way, good examples being afforded by rodent, rabbit, and mole traps. Sticky traps are widely used to assess and reduce insect numbers; for instance a number of non-flying fruit tree pests (see section 2.1.2) are controlled in this way in orchards, and houseflies are caught by this technique the world over. Artificial light is used in many parts of the world to lure insects to a pesticide or trap.

Even the hand-picking of insects from infested plants or animals, which is too costly to be considered in most highly organised enterprises, is used in certain regions of the world where labour is cheap. The vegetation cover, besides being a source of food for various pests, influences the macro- or micro-environment so that certain animals may be encouraged to nest or breed within or in its vicinity. Each vegetation type has its characteristic animal, and plant communities and the removal or alteration of the vegetation will obviously profoundly affect each of these. The removal of breeding grounds, places of refuge or alternative feeding sites by mechanised clearing or burning is used to control a number of pests, as in the control of tsetse flies (which transmit trypanosomes) in certain parts of Africa. One of the problems arising from clearing natural vegetation in many parts of the world, is the susceptibility to erosion because the elements imparting stability to the soil are removed.

A vivid illustration of a physical means of direct pest reduction is seen in certain parts of the world where the poor rural population can do nothing other than beat to death hordes of advancing locusts in an attempt to protect their crops. Other methods of physical control include the use of electric fly traps, and the flooding of land to drown certain crop pests which live in the soil. Grain may be separated from its beetle pests by sieving, and flotation techniques are used to separate light grains which may be damaged, or infested with eelworms.

Heat is widely used, even though one may fail to realise it, to kill off many potential pests in cooking. Indeed, the failure to cook meat adequately is the main reason for the perpetuation of trichinosis (see p. 44) and the broad fish tapeworm (*Diphyllobothrium latum*), the pork (*Taenia solium*) and beef (*T. saginata*) tapeworms and many other parasites which infest man. Heat is also used to sterilise utensils and storage space of various sorts which may harbour pests. In the southern U.S.A. grain is freed of insects and other pests by being heated to a critical temperature which kills them but not the grain. Hot water treatment of plants is a widely used and efficient method of freeing them of eelworms (see section 2.2.4). Again such treatment depends for its use upon the fact that provided the time-temperature tolerance relationships are known for both plant and pest, the pest can be killed by heat which is not lethal to the plant.

A number of pests are killed by low temperatures. For instance insect pests of stored products, such as the rice weevils (*Sitophilus oryzae*) and grain moths (e.g. *Sitotroga cerealella*) are killed when the grain in which they are stored is cooled. In practical terms control is achieved by opening up the store and exposing the grain when the ambient temperature is lethal to these pests. *Trichina* worms and tapeworm cysts are killed when meat is frozen and use is made of this by keeping infected or suspect meat for a prescribed period of time in cold storage.

Newer techniques of physical control, continually being brought to light, vary in their usage and efficiency. Ultrasonic vibrations of particular frequency have been used to kill grain weevils in grain though this method has not found much practical use. Ultra-violet radiation, visual light and infra-red radiation have also been examined experimentally as methods of insect control but only a few situations show practical potential. Their ultimate exploitation as control possibilities awaits the results of further research, and of developments in effective instrumentation.

Gamma and other types of radiation have also been used with considerable success to eliminate or reduce particular insect pest problems especially those of stored products (see also section 6.2 where the use of lower doses of radiation are described, which cause sterility rather than death). The economic feasibility of this technique still remains to be proven and there are a number of limitations to its use. For instance, ionising radiation treatment cannot be recommended for insect control in grain intended for use as seed since treated seed often fails to grow, but such treated seed may be used for milling and baking.

3.2 Direct Reduction by Inorganic and Plant-derived Chemical Agents

3.2.1 General considerations

Man's use of poisons for the destruction of both human and non-human enemies has a long history extending back to the remote and unrecorded discoveries of naturally occurring toxic agents in both inorganic and organic forms. In modern times poisoning has become a highly scientific procedure associated with the development of large numbers of drugs for the destruction of various pathogenic micro-organisms, anthelmintics which act on parasitic worms, selective herbicides to destroy weeds and leave valuable crops unharmed, and acaricides and insecticides to keep man's arthropod enemies in check. Many other specialised uses of toxic chemicals exist, but we are concerned solely with pest animals, and this section is therefore concerned with anti-protozoal agents, anthelmintics, insecticides, acaricides and to a lesser extent molluscicides and rodenticides.

The complexity of modern chemical control means that scientists from widely differing disciplines are involved in its problems. Chemists, engineers, toxicologists, entomologists, parasitologists, agriculturalists and many other specialists will each adopt a different approach to the problems of the use of control chemicals, and it is the purpose of this section to first examine some of the ways in which pesticides can be categorised as a basis for understanding the descriptive sections which follow.

One of the practical ways of classifying pesticides, from the point of view of the field worker, is on the basis of their mode of entry into the pest. Thus insecticides and acaricides are often described as either stomach poisons, contact poisons, systemic poisons or fumigants. Stomach poisons are applied to the food of the pest, and they are then swallowed and absorbed into the body through the gut epithelium. Contact poisons are applied to the pest itself, or to a surface across which it will travel; they are then absorbed directly through the body wall or the feet. Systemics are chemicals which are absorbed by the host on which the pest feeds, so that the host tissues are rendered poisonous. Most systemic poisons are applied to plants where they are absorbed by roots or leaves and distributed by the transpiration and translocation movements in the plants, so that the plant juices become toxic, particularly to sap-sucking insect pests. Some systemics can be introduced into animals, however, with the result that the mammalian tissue becomes toxic to the parasitic organism feeding on it. Fumigants are chemicals which act in the vapour phase and gain entry to the pest by being absorbed during respiration.

These terms are useful and will recur in the sections which follow. They are a familiar part of the vocabulary of pest control technology, but they do have certain limitations. In the first place they are primarily applicable to the control of free-living pests. Here the differences in the mode of uptake are important and readily seen. In the case of parasitic pests, however, we less frequently see the pesticide actually contacting the pest. Whether a parasitic worm absorbs a poison from its alimentary tract or through its external cuticle may well be of rather more academic than practical interest. In the field the mode of application of the chemical to the host animal is of more direct concern than its mode of entry into the parasite.

Another limitation of this classification is that many chemicals fall into more than one category. The more volatile contact insecticides may well show fumigant properties, and it is quite common for a contact insecticide which is ingested to have a stomach poison action. Lindane, for example, can readily be classified under these three headings.

The toxicologist therefore may adopt a rather different approach to the categorisation of poisons. He may largely ignore the entry of the poison into the pest, and concentrate on the effects that it produces within the body which result in the ultimate death of the pest. In other words pesticides may be classified on their mode of action rather than their mode of entry, examples being respiratory poisons (such as cyanide which inhibits essential respiratory enzymes), neurotoxic agents or nerve poisons (such as the organophosphate insecticides), and anticoagulants (rodenticides such as warfarin which prevent normal blood clotting in the rat). This type of classification again has its limitations in that the mode of action of many widely used pesticides is by no means clearly understood. Furthermore, this is a specialist approach; the farmer is not concerned with how he interferes with a pest's metabolism, his aim is merely to reduce damage as far as possible.

The most generally applicable basis for classifying pesticides is that of the chemist. Although a great diversity of chemicals is used it is possible to make broad groupings for many of them on the basis of their chemical structure, as will be evident from the succeeding sections.

3.2.2 *Characteristics of pesticides*

(a) INORGANIC COMPOUNDS

The toxic properties of many naturally occurring inorganic substances have been known for centuries, and it was natural, therefore, that when man first attempted to poison pest animals he initially turned to these compounds. Few remain important pesticides today, since the majority suffer from the disadvantage of high toxicity to man and domestic animals, often combined with a tendency to give rise to an accumulation of persistent residues. Moreover, the synthetic organic pesticides are generally far more toxic to the target animal.

Many of the inorganic pesticides therefore have a long pedigree. Greek and Roman writers commended the use of elemental sulphur for the control of a variety of insect pests. Its use still persists on a small scale, particularly in mixtures with more toxic agents, as for example in rotenone/sulphur, and DDT/sulphur dusts, where the sulphur enhances the activity of the other active ingredient. Boiled in suspension with calcium hydroxide, sulphur combines to give the fungicide lime-sulphur. This also has valuable insecticidal and acaricidal properties, but some of the studies in orchards have shown that it may actually destroy certain predators and parasites more readily than the pests which they attack, with the result that populations of scale insects, codling moth and mites have in fact shown an increase after treatment (see Chapter 7). Lime-sulphur was also used in the past as a dip for destroying skin parasites on sheep, cattle and horses, particularly the various mites giving rise to mange. Elemental sulphur has been used as a dust against lice and was at one time fed to fowls for the treatment of coccidiosis. Sulphur burns to form the pungent gas sulphur dioxide (SO_2) which is used as a fumigant in insect control, and more particularly in mammal control.

Perhaps historically the most famous or infamous poison is white arsenic, or arsenious trioxide (As_2O_3). Extensively used by the Borgias for solving political problems, it was recommended as early as 1669 as a constituent of baits for ant control. In comparatively recent years the salts, sodium arsenite and calcium arsenite, obtained by dissolving the trioxide in the appropriate alkali, have been widely used in baits for grasshoppers, locusts, ants and termites. It is interesting to note that although the vast majority of inorganic insecticides act as stomach poisons, sodium arsenite solutions can be sprayed onto locusts to function as contact poisons. When classified on the basis of their mode of action, arsenical insecticides are termed 'protoplasmic' poisons since they precipitate protein and cause tissue disintegration, particularly in the mid-gut epithelium of the insect.

The arsenical compounds which have found the greatest use as pesticides are the arsenates, and of these, lead arsenate ($PbHAsO_4$) has proved particularly valuable. It is a powder of low solubility in water, which is normally sprayed as a suspension, and which since 1892 has been used as a stomach poison against a variety of vegetable and orchard pests, particularly codling moth on apples. However, in many parts of the world the law regarding the sale of food containing traces of arsenic is justifiably very strict. This necessitates allowing a substantial amount of time to elapse between spraying and harvesting a crop. There are other disadvantages with lead arsenate treatment, such as the leaching of very persistent residues of arsenic from the crop into the soil, and, in the case of injudiciously timed sprays on fruit trees, the poisoning of bees or the contamination of honey. However, sometimes a disadvantage can be turned to good account. Many years ago it was noticed that sheep eating orchard grass contaminated with lead arsenate passed many tapeworms out with the faeces. Lead arsenate was therefore enclosed in gelatine capsules and fed to sheep where it gave reasonable control of *Moniezia expansa*. It subsequently proved effective in calves and kids and was also used against the larvae of the bot flies, *Gasterophilus* spp., parasites in the stomach of horses.

Many of the control problems for which lead arsenate was employed are now dealt with by using synthetic organic materials. Nevertheless, this chemical still finds a place in the control chemists' armoury and its occasional application, in for example an integrated control programme for orchards (see Chapter 7) may be justified. The other important arsenate is the calcium salt, which has been widely used against forest and cotton pests.

Fluorine is found in a variety of inorganic pesticides, and as sodium fluoride has been mixed in with insect baits, and with the feed of pigs to destroy the nematodes *Ascaris suum* and the stomach worm *Hyostrongylus rubidus*. It has also long been dusted onto birds for the control of feather lice. One of the disadvantages of inorganics such as sodium fluoride is their high solubility, which means that they cannot be sprayed onto plants, since their absorption results in very serious damage to the foliage. As a consequence comparatively insoluble materials such as cryolite (sodium aluminofluoride) and calcium and barium silicofluorides were employed as stomach poisons against plant pests. The mode of action of the fluorides in insects is apparently similar to that of arsenicals; they are therefore classified together as protoplasmic poisons.

One other exception to the general rule that inorganic insecticides act as stomach poisons is that of selenium. It was recognised many years ago that plants grown in soils rich in selenium carried abnormally low populations of aphids. Selenium salts were therefore watered onto the soil as a preventative measure when serious aphid attack was expected. This is an example of an inorganic poison acting as a systemic, the selenium being taken up by the roots of the plants and ultimately rendering the sap poisonous to aphids.

Copper sulphate solution has been used in a variety of control procedures in the past, but has become most widely known for the destruction of snails. Apart from the damage snails cause to crop plants, many species are important in veterinary and medical work as intermediate hosts for helminth parasites. Although copper sulphate is being replaced by new synthetic organic molluscides, its use is still fairly common.

There are many other inorganics which have been exploited as pesticides but the status of this group at the present time does not justify a prolonged discussion of numerous examples. Most current treatments against parasite, insect and rodent pests involve synthetic organic compounds. However the situation is not static. Resistant varieties arise which necessitate a return to the older treatments, an example being the use of zinc phosphide against warfarin-resistant rats. However, even the inorganics can induce resistance, as for example in the case of resistant strains of codling moth larvae which are unharmed by lead arsenate (see section 8.2.1). The restrictions imposed upon the use of persistent synthetic compounds of the chlorinated hydrocarbon group (see section 8.1) have also meant a new lease of life for some of the inorganics. This is particularly true of the domestic market where, for example, the withdrawal of aldrin for the treatment of soil pests such as cabbage root fly has forced the gardener to turn again to inorganics such as mercurous chloride, Hg_2Cl_2, (calomel) which are less toxic to the insect but which do not show the unfortunate side effects of aldrin.

(b) PLANT PRODUCTS

The long familiarity of mankind with biologically active materials has not been confined to inorganic substances. Over many thousands of years the medicine men, witches, and herbalists have discovered quite empirically the effects on animals of many chemicals contained in the leaves, fruits, roots and bark of a huge variety of plants. Plant derivatives made up a large proportion of the numerous remedies produced in the past for the ills which afflict mankind and his domestic animals, and among these concoctions were a few drugs of real therapeutic value. Many of the biologically active materials which can be extracted from plants were probably evolved as a defence mechanism against animals feeding on the plant. Certain plants have long been known to be highly poisonous to domestic animals. This chemical protection also extends to insect attack. It is not surprising therefore, that, in the past, man has utilised plants to provide specific poisons for use against insects, parasitic protozoa and worms, hunted animals and human enemies.

Some of the most biologically potent and widely known plant products are grouped together under the name 'alkaloids'. These are heterocyclic organic compounds containing nitrogen in a carbon ring structure, which derive their name from their 'alkali-like' basic properties. Here are found such potent drugs are morphine (from poppies), cocaine (coca plant), atropine (*Atropa belladonna*, or deadly night-shade) and poisons such as strychnine (*Strychnos nux-vomica*). Some of the alkaloids have proved of great value in pest control, and one of the most famous is quinine, which has long been extracted from the bark of the South American cinchona tree. Powders and infusions made from its bark are often said to be an ancient amerindian method of treating malaria, but modern authors attribute the discovery to seventeenth-century Jesuit missionaries. The technique was introduced from Peru into Spain in 1639, and the powder was known in England in the 1650s under the name of Jesuit's Root. The pure alkaloid was isolated in 1820. Quinine taken orally, is rapidly absorbed from the digestive tract and inhibits the growth and asexual reproduction of *Plasmodium* spp., the blood-dwelling protozoa which are the causative agents of malaria. The alkaloid apparently affects the respiratory enzymes of *Plasmodium*, but it may also produce undesirable effects on the human beings dosed with the drug, particularly on their hearing, vision and heart beat. Sometimes the side-effects are more serious and this has hastened the modern replacement of quinine by far more effective synthetic anti-malarial drugs.

The powerful pharmacological effects of most alkaloids makes them a promising group of chemicals in which to look for pesticidal compounds. One example of an alkaloid which has found widespread use as an insecticide is nicotine, which is extracted from the tobacco plant. It is named after John Nicot, who introduced the seeds of tobacco into Europe in 1560. Its insecticidal properties had been discovered by 1690, when tobacco infusions were used against a pear lace bug in France. Advice on the use of these infusions against a plum weevil was included in a letter sent from England to an American botanist in 1746. The alkaloid itself was isolated in 1828, and finally synthesised in 1904.

Although the recent consumption of nicotine-containing insecticides has declined as synthetic materials have appeared, it is still widely used. It is usually extracted from the coarse parts of the tobacco plant which are by-products of tobacco manufacture. Usually the free base is liberated with alkali and removed by steam distillation, but solvent extraction may be employed. The free alkaloid is quite volatile and exerts its action through the vapour entering the trachea of the insect, in other words it acts as a fumigant. However, nicotine may also act as a contact insecticide, being absorbed directly through the cuticle. It is short-lived in its free and volatile form and if more persistence is required the acidified, but less toxic forms, such as nicotine sulphate and nicotine tannate, are used. When 'fixed' in this way nicotine may be used as a stomach poison. Nicotine acts on the nervous system, causing brief hyperstimulation followed by paralysis and death. Since it acts indiscriminately, its use against insects produces risks of acute poisoning to man and domestic animals, necessitating great care at the time of application. However its volatility is such that there is no residue problem, provided that a suitable time is allowed to elapse before harvesting, this varying according to the persistence of the formulation used. One advantage nicotine shares with most other plant-derived pesticides is that it is not phytotoxic, that is to say it will not 'scorch' or otherwise damage crops to which it is applied.

In spite of their general toxicity, alkaloids can, if the dose is carefully controlled, be administered to man and domestic animals. The use of quinine and various alkaloid drugs has been mentioned above, and this biological activity has occasionally been directed against intestinal parasites. Even nicotine, in its sulphate form, was at one time mixed with copper sulphate and used against various gut-dwelling tapeworms in sheep and cattle, and against the nematode *Ascaridia galli* in poultry. An ancient remedy for tapeworms, widely used in the orient, was the administration of powdered areca nut, the fruit of an East Indian palm. Betel nut, which is commonly chewed in various parts of the Far East, is a mixture of the areca powder, betel leaves and lime, and is well known for the bright red discolouration it produces in the saliva. The activity of the alkaloid, arecoline, is very short-lived in these natural preparations and the hydrobromide salt is now normally administered instead.

The main effect of arecoline hydrobromide is directed against tapeworms and it has been commonly used in dogs to control *Taenia* spp., and *Dipylidium caninum*, but is nowadays almost exclusively used against *Echinococcus granulosus*. Arecoline hydrobromide causes relaxation of the tapeworm muscles, so that the scolex becomes detached from the intestinal wall of the dog. The effects shown by the dog itself are similar to those produced by stimulation of the parasympathetic nervous system, the rate of intestinal contraction increases, hastening the purging of the worms from the body. As with all alkaloids mammalian toxicity is a serious consideration, and arecoline causes vomiting and diarrhoea in the dog, overdosage producing convulsions and unconsciousness. In cats, the drug produces large amounts of bronchial secretion which may threaten to drown the animal, and it is not normally used to treat feline tapeworms. This again is an example of the older remedies being replaced, one by one, by safer synthetic materials.

The naturally occurring alkaloids are by no means the only plant derivatives which have found use as pesticides. Other anthelmintics which have been used include various oils extracted from plants, examples being oleoresin of aspidium, an extract of the male fern, *Dryopteris filix-mas*, and oil of Chenopodium extracted from *Chenopodium ambrosioides*, the Jerusalem oak. Oleoresin of aspidium contains a mixture of complex phenolic substances which effectively remove various species of tapeworm. It has been used against cestodes in humans but the toxic side effects it produces have rendered it virtually obsolete. These toxicity problems also apply to oil of Chenopodium or wormseed oil which has been used against strongyle nematode infections in horses and ascarid nematodes in pigs and dogs. The uncertainty of its effect and the appearance of more reliable and less toxic compounds have led to its abandonment.

In a few cases, however, plant-derived pesticides are still holding their place in modern control technology. The best examples are the mixtures of esters known as the pyrethrins and cinerins which are valuable insecticides derived from pyrethrum plants, the most important of which is *Chrysanthemum cinerariaefolium.* This species is a native of Yugoslavia and the insecticidal properties of the plant were exploited in Eastern Europe well over a century ago. Powder, from the dried flowers, was introduced into France in 1850 and the U.S.A. in 1885. In 1881 the growing plant was introduced into Japan which, over the next forty years, became the principal world supplier. In the 1920s experimental cropping began in Kenya and by the outbreak of the Second World War, Africa had overtaken Japan as the world's major source of this insecticide. It is now grown on a very wide scale in Tanzania, the Congo and particularly Kenya, and has become a major item in the economy of the region.

The mixture of active ingredients produced within the plant are commonly known collectively as the insecticide 'Pyrethrum'. It is a contact poison obtained by first drying and grinding the flower heads. Ground flower heads are still used as an insecticidal dust, but much of the insecticide remains locked away in this form. The common practice nowadays is, therefore, the less wasteful one of extracting the dried material with one of a variety of organic solvents. The extract can be sprayed as an oil-based spray, or as an emulsion with water, or in the form of an aerosol (see section 5.2.2). Dry formulations are obtained by impregnating the extracts onto dusts such as talc or ground gypsum.

The fact that pyrethrum is non-toxic to mammals results in its widespread use in food stores against many species of stored products pests. Very large quantities of the extracts, diluted with odourless kerosene, are used each year in the manufacture of fly sprays, its lack of mammalian toxicity making it also ideal for this purpose. This property further enables it to be used on crops at harvest time, but its use is restricted by its lack of persistence due to the rapidity with which it undergoes oxidation and also decomposition by light. This instability raises storage problems with pyrethrum, but deterioration may be retarded by adding antoxidants such as the phenolic compounds resorcinol and hydroquinone.

The active ingredients of pyrethrum are four esters, each formed by the combination of one of two acids with one of two ketone alcohols. The two alcohols are known as pyrethrolone and cinerolone, and the two acids are chrysanthemum monocarboxylic acid and the monomethyl ester of chrysanthemum dicarboxylic acid. When esterified with pyrethrolone these two acids give rise to the two active ingredients known as pyrethrins, and when cinerolone is the alcohol involved the esters are known as the cinerins. All four active esters are present in commercial pyrethrum and possibly partly because of this complexity of the insecticide it is very uncommon for insects to develop resistance to pyrethrum.

One of the facets of pesticide technology which is particularly well illustrated by some pyrethrum formulations is the phenomenon of synergism. Synergists are substances which are not always themselves pesticides, but which, when mixed with pesticides, greatly increase the toxicity of the formulation. There are a number of substances, for example sesamin (extracted from sesame oil) piperine and piperonyl butoxide, which show this synergistic effect when mixed with pyrethrum.

One of the most valuable properties of pyrethrum, particularly when it is used against flying domestic pests, is its ability to cause very rapid paralysis or 'knock-down'. On the basis of its mode of action it is classified as a nerve poison, since it causes brief initial nerve excitation, followed rapidly by paralysis and death, accompanied by tissue degradation in the nerve cord and ganglia. The insecticide is carried around the body dissolved in the lipoid sheaths of the nerves for which it has a great affinity. It is toxic to all cold-blooded animals with a developed nervous system, but the warm-blooded birds and mammals are resistant.

Before leaving the subject of plant derivatives brief mention should be made of the rotenoids, substances used for centuries as fish poisons in various parts of the world, which have been recognised during the last hundred years or so as insecticides. They are extracted from the roots of various leguminous plants, particularly species of *Derris* found in the Far East, and *Lonchocarpus* from South America. The main active principle, rotenone, is an exceedingly complex chemical which is extracted from the dried and ground roots with suitable solvents and used mainly as a contact insecticide, although it can also show stomach poisoning properties. The chemical is formulated in a similar manner to pyrethrum, but its somewhat greater stability has led to its having much wider use on pests of growing crops. However, associated with this stability is a small residue problem on crops and also a rather greater toxicity than pyrethrum to some species of mammals. Nevertheless it is an effective insecticide, sharing with the other plant derivatives a lack of phytotoxicity, and its use, though declining, is still very widespread. On the basis of its mode of action it can be classified as a respiratory poison in that it blocks the transfer of hydrogen from reduced coenzyme I (NADH), thereby dislocating the whole respiratory system.

Chapter Four
CONTROL OF PEST DAMAGE BY DIRECT POPULATION REDUCTION
The use of Synthetic Organic Compounds

4.1 Introductory

The previous chapter has described the use of inorganic and plant-derived poisons in pest control but at the present time they are more of historic than practical interest. In the past three decades revolutionary changes have occurred in the field of chemical control as a result of the discovery of pesticidal activity in a large number of purely synthetic chemicals. These chemicals have proved so effective that in some quarters successful pest control is regarded as synonymous with the correct choice of a synthetic pesticide. The disadvantages of this attitude will be clarified in Chapters 7 and 8, but first some of the main groups of synthetic compounds which have been used over the past fifty years will be reviewed.

4.2 The Chemicals

4.2.1 Organo-metal compounds

Some of the synthetic compounds contain toxic elements, such as arsenic, which have already been encountered in the discussion on inorganic substances, and it is with these organo-metal compounds we will deal first. In one sense at least, the organic compounds of arsenic are an appropriate starting point for a discussion of synthetic pesticides, since these were among the first compounds developed for the destruction of animal organisms causing disease in man and domestic animals, in other words the first synthetic agents used in 'chemotherapy'.

At the beginning of the twentieth century it was known that trypanosomes caused sleeping sickness, a disease rendering huge areas of Africa uninhabitable at that time. Early attempts were made to poison these parasites by dosing patients with arsenious oxide, and this was superseded by the slightly less dangerous substance atoxyl. Ehrlich showed that the structure of atoxyl was:

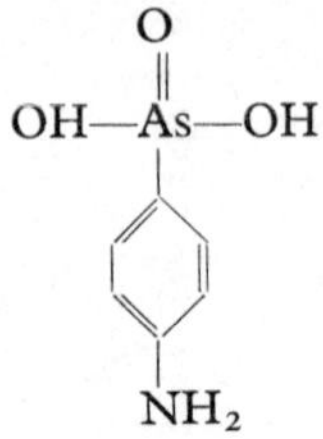

I
Atoxyl

and he set about examining derivatives in the hope of finding trypanocides of lower mammalian toxicity. Reduction produced a promising chemotherapeutic agent, arsenobenzol

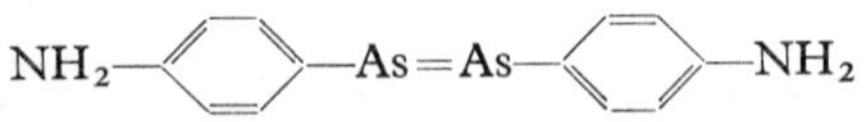

II
Arsenobenzol

and further modification of this molecule gave the famous drug salvarsan.

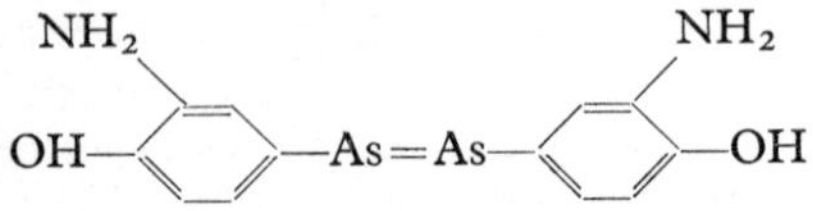

III
Salvarsan

Use of a substituted salvarsan, neosalvarsan, was the standard treatment for the spirochaete infections that cause relapsing fever and syphilis, until the discovery of antibiotics in the 1940s.

The other important reduction product of atoxyl was arsenoxide:

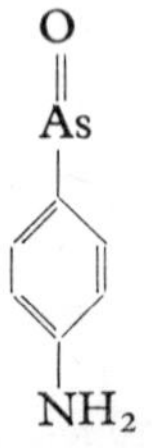

IV
Arsenoxide

If one of the hydrogen atoms in the amine group of arsenoxide is replaced by a particular complex heterocyclic ring structure, melarsen oxide is obtained.

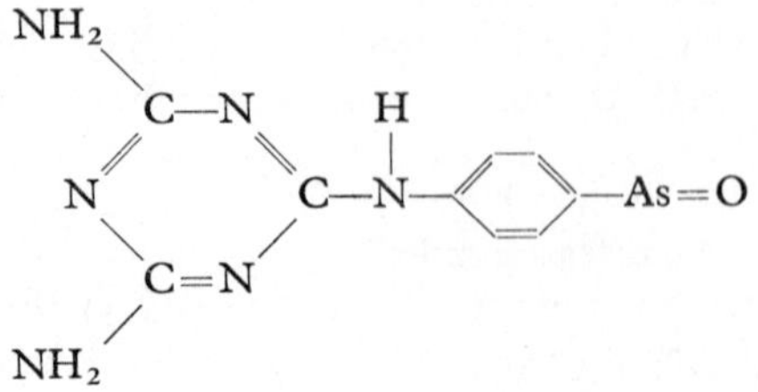

V
Melarsen oxide

This substance, discovered in 1941, forms the basis for a series of substances of very considerable trypanocidal activity which are widely used in the treatment of human sleeping sickness, the best known of these being the drug Mel B. The complete list of organic arsenicals which have been used against parasites is a considerable one. Not only spirochaetes and trypanosomes, but other organisms as well have yielded to arsenical treatment, for example the use of arsenamide against filarial nematodes. However, in many of these other fields the toxic side effects of the drugs have led to their gradual replacement. In insect control as well, organic arsenic compounds, such as the copper containing complex known as 'Paris Green', have enjoyed some popularity. Some of these compounds have an advantage over the inorganic arsenic compounds in that they show contact as well as stomach poisoning properties, but their use has not become widespread.

A second group of organo-metal poisons of proven value as pesticides are compounds containing antimony. Some of these have had a wide ranging history of use, a good example being tartar emetic, or potassium antimonyl tartrate. This compound has been used as a stomach poison against thrips and fruit flies and also in ant baits. It was one of the first chemotherapeutic agents used against trypanosomes in cattle (the more effective arsenic compounds generally being rather expensive for veterinary use) and it was also used in a now obsolete treatment of filariasis. Until recently infections of the three species of blood worms, or schistosomes, in human beings were treated by intravenous injection of tartar emetic, but the modern tendency is to use the closely related but less toxic sodium salt in place of the potassium salt. Very much more complex organic compounds of antimony such as sodium stibogluconate and hydroxystibamidine, are very extensively employed in the treatment of human diseases caused by parasitic protozoa, notably various forms of leishmaniasis.

A third group of useful organo-metal pesticides consists of compounds containing tin. For centuries metallic tin has been used to treat tapeworm infections and in recent years inorganic tin compounds have also met with some success. The natural development has been to experiment with organic tin compounds, but the majority suffer from the disadvantage of high mammalian toxicity, some being able to cause death after absorption through the skin. However one compound has been particularly widely tested, this is di-n-butyltin dilaurate

$$(C_4H_9)_2Sn(C_{11}H_{23}COO)_2 \qquad \text{VI}$$

Di-n-butyltin dilaurate

commonly known as butynorate. This showed early promise in ridding poultry of the tapeworms *Davainea proglottina* and *Raillietina cesticillus* and has been shown to be effective against various other tapeworms in several hosts.

Diphenyl- and triphenyl-tin compounds have also been used as taeniaecides and there is an interesting link here with a completely separate field of pest control, the sterilisation of insects. Derivatives of triphenyltin:

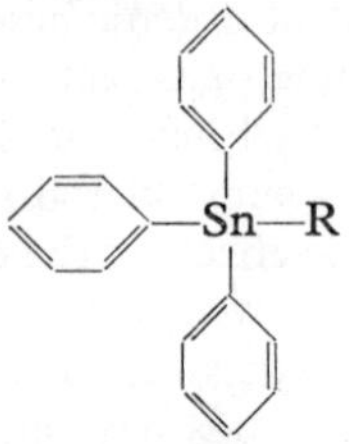

VII

Triphenyltin compounds

where R is a halide, hydroxyl, alkyl, ester or some similar group, have shown promise as chemosterilants for various insect pests. The technique of control which involves the sexual sterilisation of insects rather than killing them is dealt with in section 6.2. Triphenyltin compounds apparently are taken up with the food, contact action being ineffective normally. They affect only the female insect and the sterility induced is usually temporary, so that repeated applications may be necessary. At the present time the high mammalian toxicity of many of these organo-tin compounds is the main barrier to their further exploitation as pesticides.

4.2.2 Phenolic compounds

Some of the organo-metal compounds have been considered above as historically important in the development of the chemical control of parasites, and certain derivatives of phenol occupy a rather similar place in the development of synthetic contact insecticides. As early as 1892 one of the dinitrocresols was marketed as an insecticide in Germany, although it was not until 1925 that a really systematic study of these substances was published. The most successful was 3:5-dinitro-o-cresol, or 2 methyl 4:6 dinitrophenol,

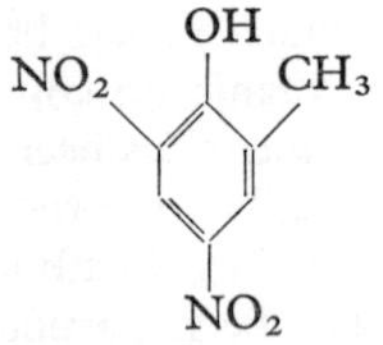

VIII

3:5-dinitro-o-cresol

also known as DNOC or DNC.

It is used as a selective herbicide as well as an insecticide, and in fact its toxicity to plant foliage is such that its main use is in 'dormant sprays' or dusts applied to fruit and occasionally deciduous forest trees in winter. For the winter spraying of fruit trees DNOC has valuable ovicidal properties, i.e. it kills the overwintering eggs of such pests as aphids and psyllids, eggs which are not susceptible to attack by many spray chemicals. Another application where phytotoxicity is not a problem has been the spraying of locust swarms in semi-desert areas.

Substitution of the methyl by a cyclohexyl (C_6H_{11}) group gives DNOCHP or 'dinex' which has the advantages of greater solubility than DNC, less phytotoxicity and a powerful acaricidal action. Certain salts of DNOCHP have lost the ovicidal action, but since they are not phytotoxic they have been used against the summer generations of red spider mite on fruit trees.

As with so many of the early contact insecticides, the main drawback with DNOC and its derivatives is their high mammalian toxicity (see Table 10). They act as respiratory poisons in all animals, but they interfere with the biochemical processes in a rather different way to that of either rotenone or cyanide. Most biological oxidations involve a chain of events by which hydrogen is removed from a molecule and ultimately combined with oxygen to form water. These 'events' are coupled to the phosphorylation reaction by which adenosine diphosphate (ADP) is converted to the energy rich triphosphate (ATP). The effect of compounds such as DNOC is to uncouple the oxidation from the phosphorylation. The reaction by which ATP is synthesised from ADP is blocked but the oxidative reactions proceed independently. In fact, one symptom of poisoning is an elevated oxygen consumption; presumably because the supply of ADP is no longer a limiting factor on the rate at which the oxidation takes place.

The pesticidal use of phenolic compounds has not been entirely confined to insect and mite control. At one time substances such as thymol and β-naphthol were employed as rather dangerous and not very effective anthelmintics. More popular, and more useful, was hexylresorcinol

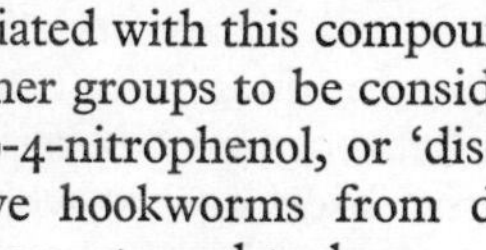

IX
Hexylresorcinol

used against *Ascaris lumbricoides*, and against hookworms in man and dogs, and also against the whipworm *Trichuris trichiura* in the human colon. However, the toxic hazards associated with this compound have led to its replacement by anthelmintics of other groups to be considered later. More effective is the compound 2:6-diiodo-4-nitrophenol, or 'disophenol', which has been successfully used to remove hookworms from dogs, and the gapeworm (*Syngamus trachea*) from pheasants and turkeys.

4.2.3 Hydrocarbons

Simple hydrocarbons have been put to a variety of pesticidal uses during the last few decades. Antimalarial activity has been found in benzene and cyclohexane derivatives, toluene has been used to remove ascarid nematodes and the larvae of the bot fly (*Gasterophilus intestinalis*) from horses, and the same compound is effective against a variety of roundworms in cattle. Naphthalene has a long history of use for such problems as the protection of timber from boring insects and the repulsion of lice from man and domestic animals; it has also been employed as a fumigant against greenhouse red spider mite. However the major pesticidal importance of the hydrocarbons lies not in these simple compounds, but in the much more complex mixtures derived from naturally occurring petroleum, or from the destructive distillation of coal. These sources provide many important solvents for pesticide formulations, and also hydrocarbon oils which are useful insecticides in their own right.

One of the by-products of coal gas manufacture is coal tar, which on distillation yields oils which include compounds used as insecticides for more than fifty years. These 'tar oil distillates' are derived from the higher boiling point anthracene oil and creosote oil fractions and were for many years the most important winter, or dormant sprays for use on fruit trees. Water emulsions of tar oil are ovicidal to aphid and psyllid eggs but their very high phytotoxicity precludes their use on plants bearing foliage.

Oils derived from petroleum have been used for more than a century in insect control. The distillation of petroleum yields a variety of fractions classified on the basis of their boiling point. Among the most volatile fractions is gasoline (or petrol) which is followed by kerosene, this being the first petroleum derivative ever to be used as an insecticide (in 1865), the target organisms being citrus pests. The high volatility and phytoxicity of kerosene has confined its present day use to fly sprays, and it is from the heavier oil fractions of higher boiling point that most petroleum insecticides are now derived. Largely unrefined petroleum oils may be used as dormant sprays, or even early in the development of the foliage on fruit trees. They kill overwintering eggs of capsids and red spiders, but not aphids or psyllids. Thus, the old winter treatment of orchards for these four groups of pests was a tar oil spray followed by a petroleum oil spray. The modern tendency is to incorporate DNC into the petroleum oil spray and destroy all four groups of eggs in one application, the tar oil application having been abandoned. Much more highly refined petroleum oils have far less phytoxicity and can be applied as summer oil sprays which are insecticidal in their own right, and into which other insecticides and fungicides can be incorporated.

The mode of action of hydrocarbon oils varies according to the volatility of the fraction used and the insect against which it is employed. Generally they appear either to block the respiratory openings and suffocate the egg or insect, or they are absorbed through the cuticular lipids allowing certain of their constituents to destroy the nervous tissue.

4.2.4 The halogenated hydrocarbons

The possible scope and value of using hydrocarbons as pesticides is enormously increased by the introduction of halogen atoms, particularly chlorine and bromine, into their molecules. Even the simplest aliphatic paraffins such as methane (CH_4) and ethane (C_2H_6) are transformed into various extremely valuable pesticides by substitution with halogen atoms. Methyl bromide

(CH_3Br) is a highly penetrative fumigant, its wide range of uses including the destruction of stored products pests in ships, mills and warehouses, the fumigation of nursery stock to kill insect eggs as well as active stages, the treatment of seeds and the fumigation of soil to destroy insect and eelworm pests. It has also been used for the eradication of termites from houses, and during the Second World War methyl bromide fumigation in airtight bags was one of the standard methods for delousing uniforms in the U.S. army. Methyl bromide is an irritant which acts as a protoplasmic poison, proteins apparently being precipitated by the acids released in contaminated tissues. It also inhibits the respiratory enzyme succinic dehydrogenase, a property which may be related to the brain damage associated with methyl bromide poisoning in human beings. In the hands of competent operators it is not a dangerous fumigant but its odour is difficult to detect and care must be exercised in its use.

It is possible to replace each of the hydrogen atoms in a simple hydrocarbon by halogen atoms. An example is carbon tetrachloride (CCl_4) which, although it has low insecticidal activity, is used as a fumigant in, for example, domestic dwellings since it has the advantage over many other fumigants of being non-inflammable. Because of this property it is more usually used in mixtures with other more toxic fumigants where it serves the purpose of reducing the risk of fire or explosion.

The pesticidal properties of carbon tetrachloride have been widely exploited in the control of parasites of domestic animals. Sheep infected with the liver fluke, *Fasciola hepatica* are treated with CCl_4 which is used either enclosed within a capsule, or as an injection, or as a mixture in liquid paraffin. Overdosing can result in fatty degeneration of the liver and in kidney damage, these toxic side effects being accentuated by malnutrition or undernutrition. Cattle are even more susceptible to these effects and are usually treated with the analogue hexachloroethane (C_2Cl_6), which, like CCl_4, is effective only against mature flukes, but is relatively safer. Not only flukes are eliminated by these substances; carbon tetrachloride has been used against stomach worms and hookworms in various host species, against strongyle nematodes in horses, and against ascarids in all domestic animals except pigs (which happen to be particularly sensitive to CCl_4 poisoning). In the case of hookworms, this treatment was superseded by the use of tetrachloroethylene ($C_2H_2Cl_4$) but this in its turn has given way to more recent synthetics considered later. At the present time, only in the case of fluke infestations in sheep does the use of CCl_4 survive as a practical control measure.

Many further examples of the pesticidal uses of this class of compounds could be given. The non-inflammable dry-cleaning fluid trichloroethylene (C_2HCl_3) has found some use as a substitute for CCl_4 in insect fumigation. Ethylene dibromide ($C_2H_4Br_2$) has gained popularity as a fumigant for grain pests and eelworms and the mixed halogen molecule ethylene chlorobromide

(C_2H_4BrCl) has been shown to be a suitable fumigant for use against fruit flies. Of the higher molecular weight derivatives, perhaps the best known is the mixture of propylene chlorides which is usually referred to as D-D mixture and is a valuable fumigant for soil application in the control of eelworms.

It is of course possible to introduce halogens as substituents into a simple hydrocarbon molecule, not as single atoms, but as part of a more complex substituent group. This is the case, for example, in the halogenated diphenyl methanes, one important derivative of which is dichlorophen.

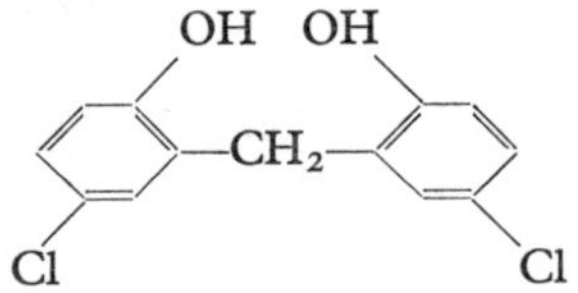

X

Dichlorophen

Dichlorophen has been widely used for treating the following tapeworm infestations, *Taenia saginata* in human beings, and *Taenia* spp. and *Dipylidium caninum* in cats and dogs. Unfortunately it has little effect on the tapeworm *Echinococcus granulosus*, for which the more toxic arecoline hydrobromide (see 3.2.2(b)) remains the standard treatment.

If more chlorine atoms are introduced into the molecule one obtains the structure of hexachlorophene.

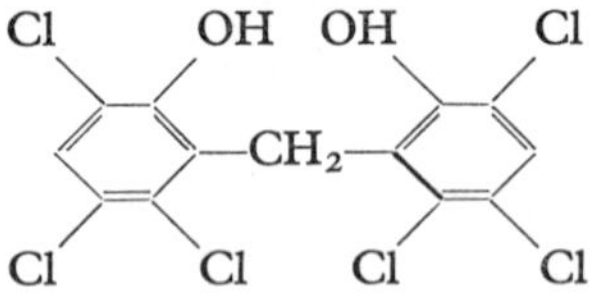

XI

Hexachlorophene

This has been used against the tapeworm *Raillietina cesticillus* in poultry, but is mainly of value as an alternative to CCl_4 in the treatment of fluke infestations. Its administration is now one of the routine treatments for *Fasciola hepatica* and *Paramphistomum* sp. in sheep and it has met with success in the treatment of fascioliasis in cattle.

Both of these compounds were considered as being derived from methane by the substitution of halogen attached to a benzene ring. This idea can be extended to ethane, the next member of the paraffin series. Substitution of the two chlorophenyl groups into the molecule of trichloroethane gives

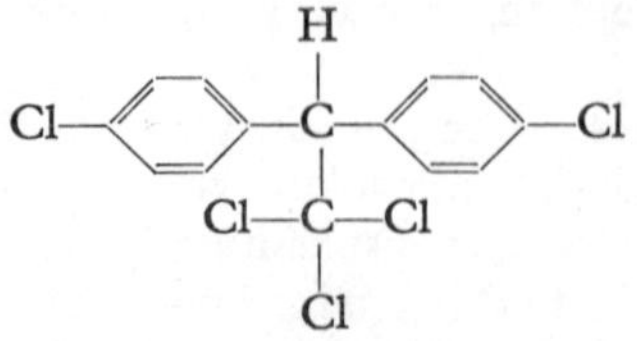

XII

αα-bis-(p-chlorophenyl) βββ trichloroethane

more loosely called dichlorodiphenyltrichloroethane, which is normally abbreviated to DDT. This molecule was first synthesised in 1874 but it was not recognised as an insecticide until 1939 when, as a result of an investigation by a team of Swiss workers, Muller rediscovered DDT and demonstrated its potent contact action in a field infested with Colorado beetle. The Swiss patent was taken out in 1940 but wartime conditions restricted the widespread publicising of the discovery. The Western allies became aware of the importance of DDT in 1942, the supplies of pyrethrum from Japan, and derris from the East Indies no longer being available to them. The chemical was valuable in that it effectively destroyed a wide range of insect pests on growing crops and in stored grain, and perhaps most important of all under wartime conditions, most of the insect vectors of human disease. In the majority of the wars in human history disease has caused at least as many casualties as the actual conflict, this being particularly true of typhus, carried by the louse *Pediculus humanus corporis*. In 1943 an outbreak of typhus occurred in Naples, but was brought under control in an unprecedented brief period of time, as a result of the treatment of 2½ million people with DDT powder (see Plate 6B). Since the war an enormous range of DDT formulations has been used to deal with vast numbers of insect pest problems all over the world. The economic and health benefits have without doubt been tremendous, but problems have arisen, and the consequences in terms of population growth, ecological disturbances, and residue and resistance problems are considered elsewhere in this book.

DDT is readily soluble in many of the common organic solvents, it is extremely stable and affords long term protection from insect attack. These properties were originally hailed as great advantages but the modern attitude is that the ready solubility, particularly in animal fats, and long persistence of the residues must inevitably lead to the ultimate abandonment of the widespread application of this pesticide (see section 8.1).

DDT acts primarily as a contact poison, but it is also a potent stomach poison. In spite of several investigations during the period of almost thirty years since the discovery of its insecticidal properties, a complete picture of the reasons for all of the effects it produces in poisoned insects has not yet been produced. The symptoms of poisoning such as hyperactivity, tremors, loss of water and fluid faeces from the gut, depletion of carbohydrate reserves, and various other forms of metabolic malfunction provide clues to the metabolic and neuroendocrine dislocations within the insect. Many of these consequences of poisoning could cause the ultimate death of the insect but one of the difficulties is to disentangle the primary effect of DDT on the insect from all of the secondary consequences which may result. It now seems clear that DDT is a nerve poison which probably acts first on the peripheral nerves of the insect, interfering with conduction perhaps by affecting the permeability of the nerve to potassium. There is a suggestion that as a result of these dislocations in the peripheral activity, the central nervous system is stimulated to produce a toxin which ultimately evokes many of the secondary symptoms.

Many analogues of DDT have been synthesised and several show insecticidal activity. Of these, three compounds have been particularly exploited commercially, methoxychlor, DDD and DFDT. In methoxychlor the Cl atoms in the 'para' positions are replaced by methoxy groups.

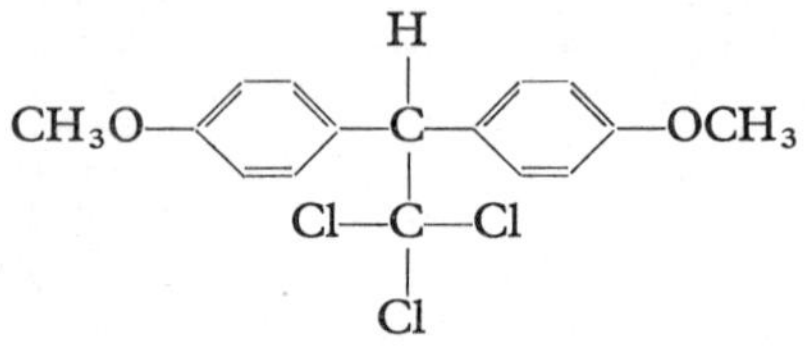

XIII

Methoxychlor

DDD (also known as TDE) is the dichloroethane analogue of DDT.

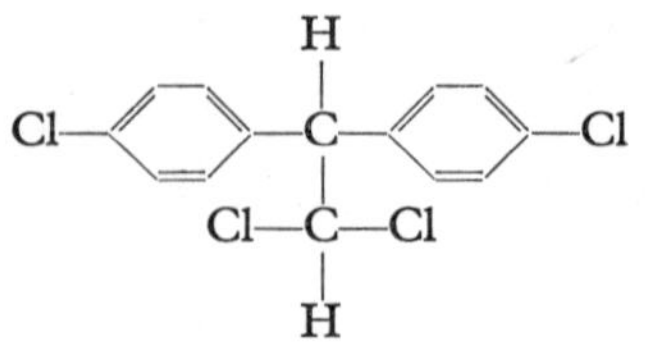

XIV

DDD

DFDT is the parafluoro analogue of DDT.

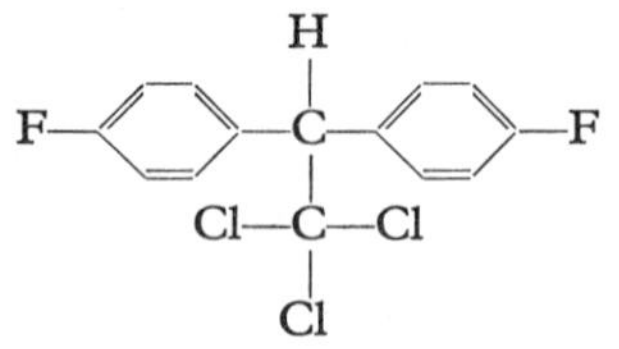

XV

DFDT

Methoxychlor has a very low mammalian toxicity and is sometimes preferred to DDT for use on food crops. Its particular value has been in biting fly control on livestock where the solubility of DDT in body fat and milk is a distinct disadvantage. DDD has a similar advantage of lower acute mammalian toxicity than DDT. It has been used against arthropod vectors and against certain crop pests such as the tobacco hornworm (*Protoparce sexta*) and various species of tortricid moths, which are much more susceptible to DDD than to DDT. DFDT has been produced commercially in Germany and, although generally less insecticidal than DDT, it has the advantage of higher volatility, which conveys some fumigant properties and results in far less persistent residues.

The phytotoxicity of DDT and its analogues is not normally very great but none of them can be used on plants of the family Cucurbitaceae (melons, cucumbers, etc.).

Instead of substituting halogen-containing groups for the hydrogen atoms of alphatic paraffins, they can be incorporated into unsaturated aromatic cyclic hydrocarbons to give an alternative range of pesticides. For example, one of the products of the introduction of two chlorine atoms into the benzene ring structure is paradichlorobenzene.

XVI

Paradichlorobenzene

Paradichlorobenzene is a crystalline solid which readily gives off a fumigant vapour. Together with naphthalene it is a popular constituent of moth balls, its efficiency depending upon its being enclosed in a sufficiently air-tight atmosphere for insecticidal concentrations to be attained. It also has repellent properties and is widely used in museum boxes and insect collections to prevent the depredations by pests such as booklice. Its repellent properties have led to its recommendation for use in the quite different pest problem of preventing bats from roosting in the roofs of domestic dwellings. Several related benzene derivatives such as orthodichlorobenzene, parachloroiodobenzene and parachlorobromobenzene have been used as insecticides of minor importance.

This general group of chemicals was being examined in Great Britain in 1942 in the hope that it would provide repellents and insecticides for protecting the valuable wartime crops. It was found that considerable insecticidal activity resided in the mixture of isomers obtained when benzene was chlorinated in the presence of light. This insecticide, 1.2.3.4.5.6.-hexachlorocyclohexane, or benzene hexachloride, is usually given the abbreviated name B.H.C. ($C_6H_6Cl_6$).

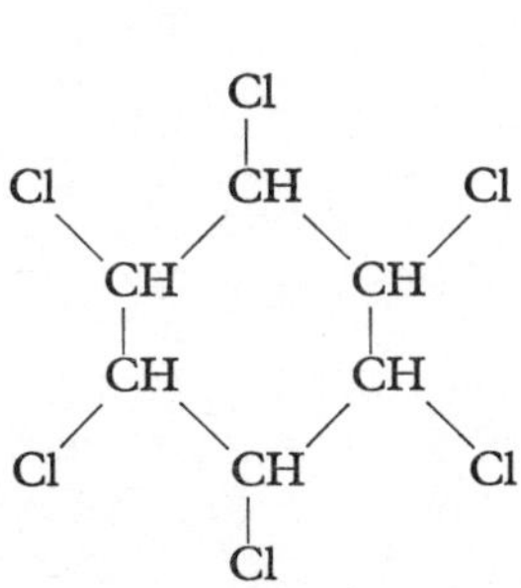

XVII

BHC

At the time of its discovery it was not known that similar findings were being made in occupied France. Further investigations in Britain showed that of the mixture of structural isomers which made up the crude BHC preparations, one, the γ-isomer, was responsible for much of the insecticidal activity. Purified preparations of this isomer were subsequently marketed under the name of lindane.

Unlike DDT, both BHC and the γ-isomer lindane are appreciably volatile at normal temperatures and therefore show fumigant action in addition to their stomach and contact poisoning properties. They are therefore frequently more effective than DDT against insects living in distorted or very convoluted foliage as occurs, for example, in the case of weevil and aphid attack on some plants. This volatility also reduces the residue problem, these insecticides usually being somewhat less persistent than DDT. BHC and lindane have been used against a very large range of insect pests on both crops and domesticated mammals. BHC is particularly useful against sawflies, bark beetles and as a soil insecticide to control chafers, wireworm and rootfly larvae.

The main disadvantage of commercial BHC preparations is the possibility of affecting the crop to which they are applied. If careful attention is not given to the choice of formulation (see section 5.1), there are many crop plants which will readily show damage, or 'scorch', as a result of BHC application. Furthermore several vegetable and fruit species are liable to pick up an unpleasant odour and/or taste from BHC applications. This 'taint' may survive, and even be accentuated by cooking or canning processes. Lindane is far less phytotoxic, being suitable for safe use on plants which are readily damaged by BHC mixtures, and furthermore, the problem of taint is very markedly reduced in the case of lindane, although its use is still avoided on vegetables and fruit destined for canning.

The mode of action of lindane appears to be similar to that of DDT, but again it is not yet clearly understood. Lindane also appears to be a nerve poison, and it is suggested that, as in the case of DDT, the primary effect is on the permeability of the nerve membrane. One hypothesis is that the effect is a physical one in the sense that the shape of the insecticide molecule distorts the molecular structure of the nerve membrane, for which it has an affinity. Studies on the behaviour of poisoned insects show small differences in symptoms if BHC is applied rather than DDT. These studies confirm that the site affected is in the central nervous system, unlike DDT which affects the peripheral and sensory nerves initially (see above).

The effect on mammals of accidental ingestion of these chemicals is to produce enlargement, degeneration and necrosis of the liver and kidneys, and haemorrhages in the heart and lungs. However, the amount required to induce these effects is very large. The acute toxicity of BHC is many times less even than that of DDT, and most of this toxicity resides in the γ isomer. Lindane therefore is much more toxic to mammals than BHC, but even lindane is slightly less toxic to laboratory rats than DDT (see Table 10) and can therefore be regarded as one of the less dangerous insecticides (for further discussion of mammalian toxicity see p. 94).

As in the case of the alphatic compounds, the chlorine atoms can form part of a substituent group, rather than being attached directly to the benzene ring. One example is the introduction of two fully chlorinated methyl groups into the ring to give parabistrichloromethylbenzene.

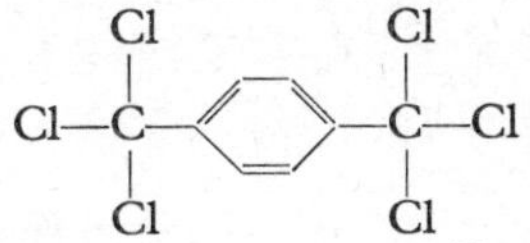

XVIII

Parabistrichloromethylbenzene

This compound is a relatively new arrival on the pesticide scene and under the name 'hetol' has shown considerable promise in the treatment of fasciolíasis in sheep and cattle. The dosing of cattle with 'hetol' is purported to provide far more effective elimination of mature fluke than hexachloroethane (see above) with apparently enormously reduced risks of toxic side effects.

The remaining two groups of halogenated hydrocarbon insecticides are compounds of higher molecular weight known as the chlorinated terpenes and cyclodienes. The terpene hydrocarbons are cyclic molecules of empirical formula $C_{10}H_{16}$, the oxygenated derivatives of which are known as camphors. The mildly insecticidal and repellent properties of natural camphor, derived from a Formosan tree, have been exploited for very many years. One of the intermediates in the commercial synthesis of camphor is a substance, camphene, which will react with chlorine to produce a mixture of chlorinated terpenes marketed under the name toxaphene. Toxaphene has proved valuable in a diversity of pest problems such as the control of cattle ticks, the control of boll-weevils and other cotton pests, and the elimination of grasshoppers on certain crops. A similar mixture of chlorinated terpenes, known as strobane, has been incorporated into domestic pesticides for clothes moth and carpet beetle control.

The closely related cyclodiene insecticides originated in 1945 with the discovery of the toxicity of a substance $C_{10}H_6Cl_8$ with the structural formula

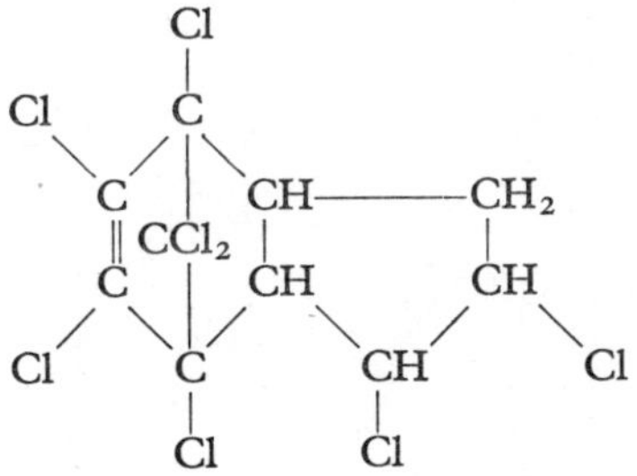

XIX

Chlordane

which subsequently became known as chlordane. This compound proved useful in the control of grasshoppers, cockroaches and ants, and also as a soil insecticide for control of wireworm and chafer infestations. Technical chlordane contains related derivatives as contaminants and one of these, which has the structural formula

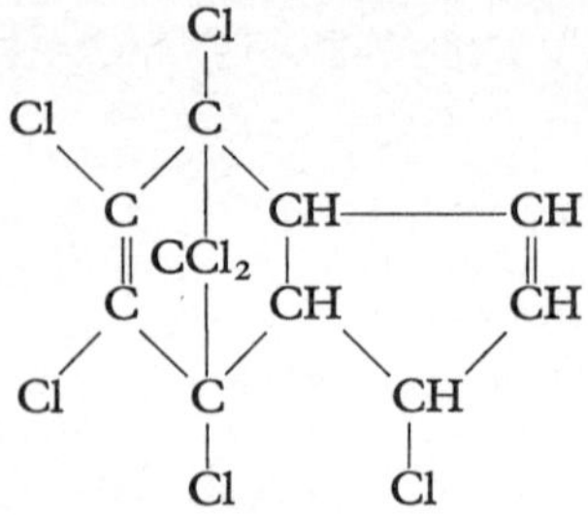

XX

Heptachlor

is a potent insecticide in its own right, marketed under the name, heptachlor. It is a powerful stomach and contact poison with some fumigant properties, which has been put to a similar range of uses as has chlordane against the more heavily sclerotised insects and soil dwelling pests.

Further searches for insecticides among the cyclodienes were rewarded in 1949 with the discovery of the toxic properties of the substance which came to be known as aldrin.

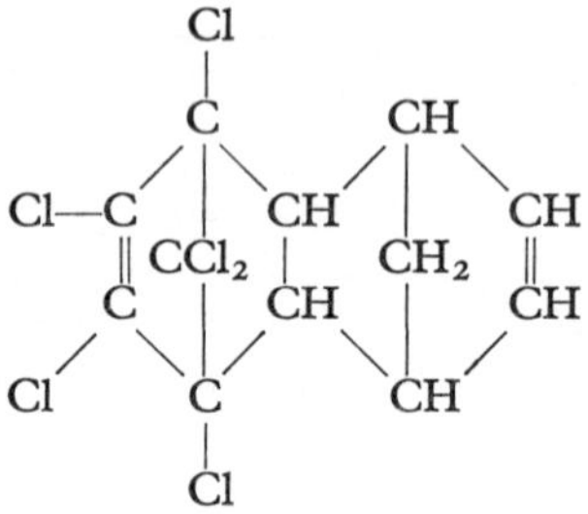

XXI

Aldrin

At the same time the insecticidal properties of the epoxide of aldrin were recognised and this substance was given the name dieldrin.

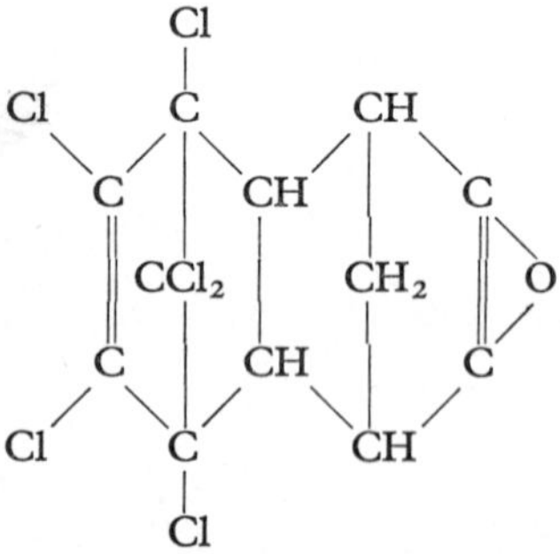

XXII

Dieldrin

It is of course possible to alter the spatial configuration of the atoms within these molecules and produce isomers which are represented by the same structural formulae as those shown above. Isodrin, which was discovered in

1950, is one such stereoisomer of aldrin. The epoxide of isodrin, which is of course a stereoisomer of dieldrin, is the well-known insecticide endrin.

Aldrin and dieldrin have been used very extensively against pests of crops and animals. They were the main insecticides used in soil treatment during the 1950s, their contact effect persisting for months, and in some cases, for several years after application. They are both highly toxic to insects, dieldrin generally being much more so than aldrin, and the application rates are usually very low. There is some evidence that aldrin may be oxidised to dieldrin, both within the soil and within insects affected by this poison. The cyclodienes are nerve poisons, their primary mode of action on the central nervous system apparently being very similar, if not identical, to that of BHC and lindane, described above.

The mammalian toxicity of the cyclodienes is generally higher than that of DDT, and endrin in particular is quite a dangerous poison if misused (see Table 10). However, the acute toxicity has proved far less of a problem than the long term effects induced by the accumulation of cyclodiene residues. The serious problems that have arisen, and the legislation which has restricted the use of these insecticides are discussed elsewhere (section 8.1).

4.2.5 The organophosphorus compounds

Many of the most valuable modern pesticides are synthetic organic compounds containing phosphorus in their molecules. Compounds of this organophosphorus group have been known to chemists for more than sixty years, but systematic investigations of their toxicity did not begin until 1934. This work was carried out in Germany by a team led by Gerhard Schrader, and several valuable pesticides were discovered. However, the rest of the world was unaware of these discoveries until their publication in 1947, the reason being that organophosphorus compounds of very high mammalian toxicity were developed for military use. These substances could be applied in very low concentrations as mists, and since they caused lethal dislocations in the mammalian central nervous system, they came to be known as nerve gases. After absorption through the human skin, death may follow extremely rapidly, and large quantities were being produced in Germany at the end of the war. A veil of secrecy also covered parallel researches carried out in Cambridge, where the toxic properties of the substance known as diisopropyl fluorophosphonate or DFP, were singled out for particularly detailed investigation.

In the post-war period, 'the swords have been converted into ploughshares' and the main line of development has been towards enhancing the desirable insecticidal properties of the group, while at the same time reducing to the greatest possible extent the very high mammalian toxicity which characterised the first insecticides produced by Schrader and his colleagues.

The chemical structure of the organophosphorus pesticides is perhaps best understood by considering them as substituted phosphoric acids. The majority can be related to orthophosphoric acid, H_3PO_4, which can be represented as

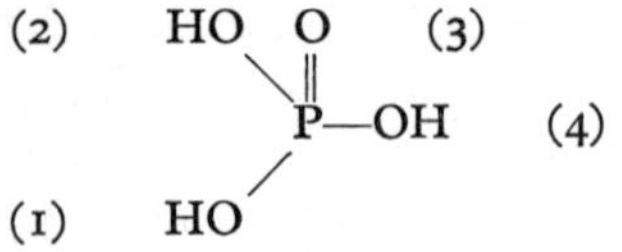

XXIII

Orthophosphoric acid

In most of the organophosphorus pesticides the hydroxyl groups (1) and (2) are replaced by one of three groups, a dimethylamino group, $(CH_3)_2N$—, a methoxy group, CH_3O—, or an ethoxy group, C_2H_2O—. Examples of these three types of organophosphates are:

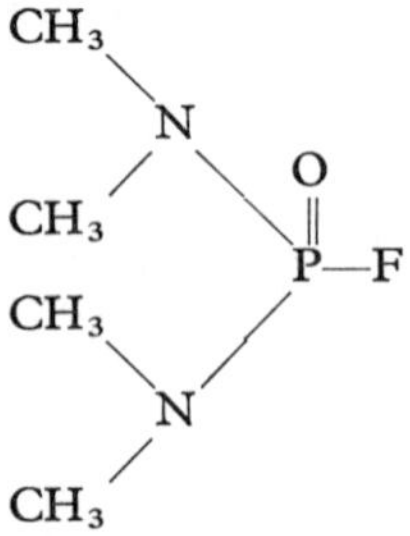

XXIV

Dimefox

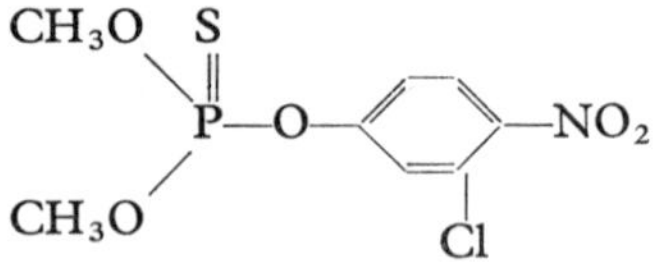

XXV

Chlorthion

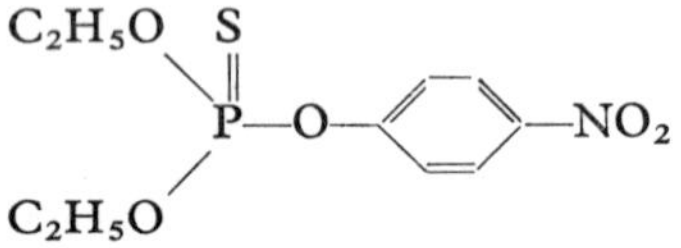

XXVI

Parathion

These compounds also illustrate the other structural features of organophosphorus pesticides. The oxygen atom (labelled (3)) in the orthophosphoric acid diagram (XXIII) may remain, as in dimefox (XXIV), but is frequently replaced by a sulphur atom, as in chlorthion (XXV) and parathion (XXVI). However, it is the replacement of the hydroxyl group numbered (4) in diagram (XXIII) which offers most scope to pesticide chemists. The very wide range of organophosphorus compounds now available is due to the great diversity of possible substituents at this point in the molecule. Usually the substituent group is derived from an acid, e.g. hydrofluoric, HF, in the case of dimefox, and paranitrophenol, 

in the case of parathion, but even this generalisation does not always hold true.

The range of organophosphorus compounds is further increased by the fact that orthophosphoric acid is not the only phosphoric acid whose derivatives have pesticidal properties. Certain very active compounds can be similarly related to pyrophosphoric acid, $H_4P_2O_7$.

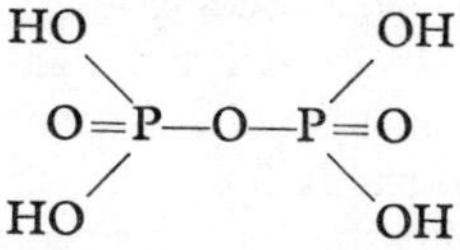

XXVII

Pyrophosphoric acid

The four hydroxyl groups can be replaced by the same alkoxy or dimethylamino groups that formed substituents at positions (1) and (2) in the orthophosphoric acid molecule, examples being TEPP (tetraethylpyrophosphate)

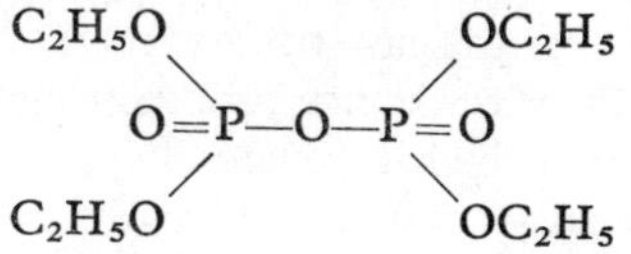

XXVIII

TEPP

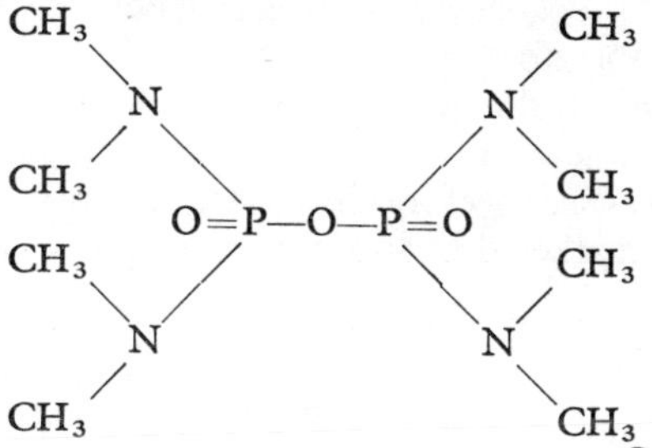

XXIX

Schradan (Pestox III)

Similarly the two oxygen atoms, connected by double bonds to the phosphorus atoms, can, as in the orthophosphates, be replaced by sulphur, e.g.:

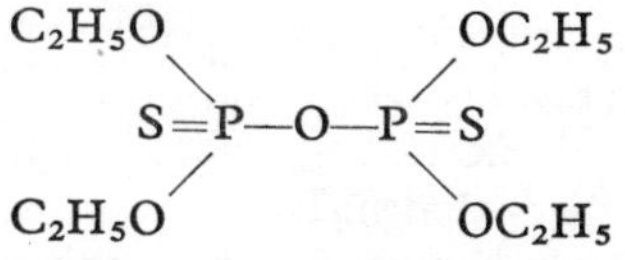

XXX

Sulfotepp (dithion)

Several of the above compounds were among the wartime discoveries of Schrader and his team. Dimefox (XXIV) was made in 1940 and parathion (XXVI) in 1944. Both found use as insecticides, but both suffered from the disadvantage of very high mammalian toxicity. They are contact poisons, and plants sprayed with parathion retain an active residual deposit on the leaf surface for up to a fortnight. On the other hand other organophosphates do not leave even these short term residues, TEPP, which was first marketed in Germany in 1944, is very rapidly hydrolysed to harmless derivatives and was therefore used on crops immediately before harvesting in spite of its high mammalian toxicity (see Table 10), its main value being to replace nicotine in aphid control. Apart from this contact action, other organophosphates have shown extremely valuable systemic properties. These were first noted in the compound discovered by Schrader in 1942, which subsequently came to be known as schradan (XXIX). Other compounds, which the plant could similarly translocate, eventually appeared, and their value in the control of sap-sucking insects is very considerable. Systemics usually enter the sprayed plant and leave very little residual contact activity on the surface, with the result that beneficial pollinating and predaceous insects are relatively unharmed, but sap-sucking insects are readily killed. The major use of these chemicals is therefore in the control of these piercing and sucking insects, many of which are incriminated as the transmitters of plant viruses.

A very large number of organophosphorus compounds have been 'screened' as potential insecticides in the last twenty years. Interest in the group has grown, partly as a result of resistance problems and toxic hazards associated with the widespread use of the halogenated hydrocarbons. Organophosphates provide no permanent answer to the resistance problem (section 8.2.1) but they are generally less persistent than organochlorine compounds, and the residue hazards are therefore extremely low.

As already mentioned, the main problem of the pesticide chemist has been the reduction of the acute mammalian toxicity of these substances, and a great deal of progress has been made. One of the major early successes was the compound first marketed in 1950 under the name malathion.

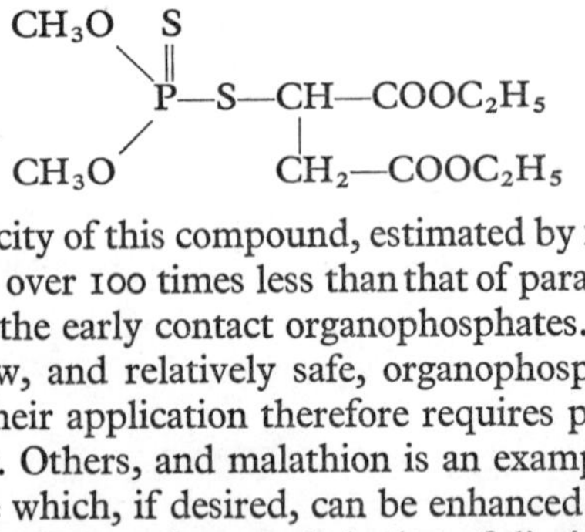

XXXI
Malathion

The mammalian toxicity of this compound, estimated by feeding to laboratory rats (see Table 10) is over 100 times less than that of parathion, which was the most widely used of the early contact organophosphates.

Some of these new, and relatively safe, organophosphates have virtually no persistence and their application therefore requires precise timing if they are to reach the pest. Others, and malathion is an example, have a moderate degree of persistence which, if desired, can be enhanced by suitable formulation (section 5.1). Another good example is that of diazinon

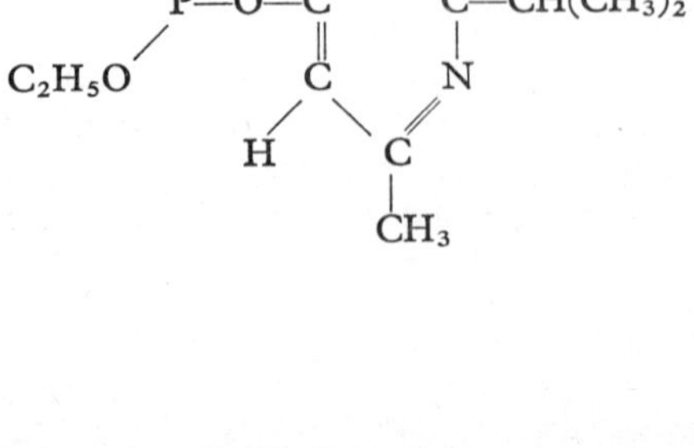

XXXII
Diazinon

which is being used in granule formulations for the control of soil pests previously attacked with cyclodienes. The activity of the diazinon granules persists for sufficient time to protect the plant from soil-dwelling larvae at the susceptible stages in the plant's growth, but there is no problem of heavy soil residues, months, or even years after application.

Parallel reductions in mammalian toxicity can be seen in many of the modern systemic poisons, one example being rogor, or dimethoate, which was introduced in 1956.

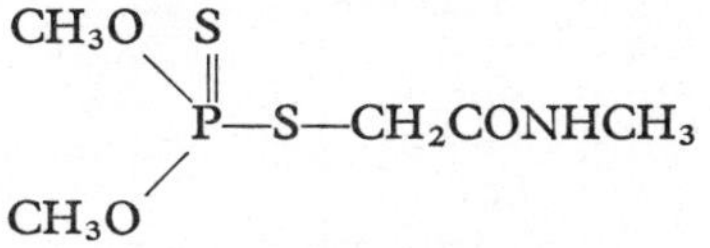

XXXIII

Rogor

Rogor has about one tenth the oral toxicity to rats of schradan, the first systemic insecticide.

The mammalian toxicities of several of the compounds mentioned throughout this chapter have been compared in Table 10. One cannot of course assume

TABLE 10 MAMMALIAN TOXICITIES OF SELECTED PESTICIDES
(All toxicities are acute oral doses for the laboratory rat; given as milligrammes of poison per kilogramme of rat)

Class of Compound	*Name*	*Formula number in text*	*LD_{50} mg/Kg*
Inorganic	Lead arsenate	—	10–50
Inorganic	Calcium arsenate	—	35–100
Inorganic	Mercuric chloride	—	37
Plant Product	Nicotine	—	50–60
Plant Product	Pyrethrum	—	800
Plant Product	Ryania	—	1200
Phenolic	DNOC	VIII	40–65
Chlorinated Hydrocarbon	DDT	XII	113
Chlorinated Hydrocarbon	Methoxychlor	XIII	6000
Chlorinated Hydrocarbon	TDE	XIV	3400
Chlorinated Hydrocarbon	γ-BHC	XVII	125
Chlorinated Hydrocarbon	Chlordane	XIX	283
Chlorinated Hydrocarbon	Heptachlor	XX	100
Chlorinated Hydrocarbon	Aldrin	XXI	67
Chlorinated Hydrocarbon	Dieldrin	XXII	46
Chlorinated Hydrocarbon	Endrin	—	17·8
Organophosphate	Dimefox	XXIV	5
Organophosphate	TEPP	XXVIII	0·5
Organophosphate	Sulfotep	XXX	5
Organophosphate	Parathion	XXVI	6·4
Organophosphate	Methyl-parathion	XXXIV	12–16

Class of Compound	*Name*	*Formula number in text*	*LD_{50} mg/Kg*
Organophosphate	Schradan	XXIX	13·5–35·5
Organophosphate	Diazinon	XXXII	100–150
Organophosphate	Rogor	XXXIII	245
Organophosphate	Chlorthion	XXV	625–1500
Organophosphate	Ronnel	XXXV	1740
Organophosphate	Malathion	XXXI	1400–1900
Carbamate	Isolan	XXXVIII	12
Carbamate	Sevin	XXXIX	500–700
Aziridine	TEPA	XLII	37
Aziridine	Metepa	—	136
Organic Sulphur	Sulphenone	LXV	>2000
Organic Sulphur	Tedion	LXVI	5000–14 700
Organic Sulphur	Chlorfenson	LXVII	2000–2050
Organic Sulphur	Chlorbenside	LXVIII	2000–10 000

The toxicity in each case is expressed as the 50% lethal dose, or LD_{50}, which is the dose required to kill, on average, 50% of a batch of experimental animals.

that the toxicity levels demonstrated in laboratory rats apply equally to man and all other mammals. There is substantial variation from one species to another, but nevertheless this is a useful way of comparing chemicals. The relationships between chemical structure and toxicity are complex and incompletely understood, and the interesting studies that are being made in this field cannot be included within the scope of this book. One minor example of the type of observation that can be made is that a comparison of the organophosphorus compounds listed in the table suggests that the introduction of chlorine into the molecule may be a factor in the reduction of mammalian toxicity. The introduction of a chlorine atom into the structure of methylparathion

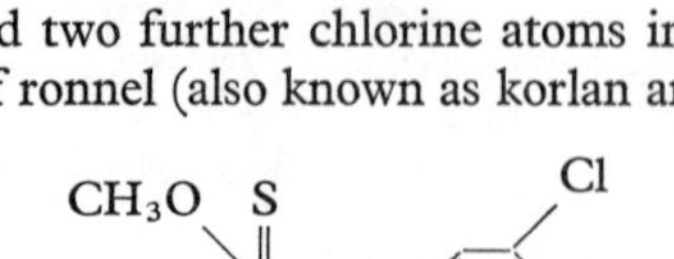

XXXIV

Methylparathion

transforms it into that of chlorthion (XXV) and the decrease in toxicity is shown by the increase in rat LD_{50} which can be seen in the table. If the nitro group is removed, and two further chlorine atoms introduced, we have the molecular structure of ronnel (also known as korlan and fenchlorphos),

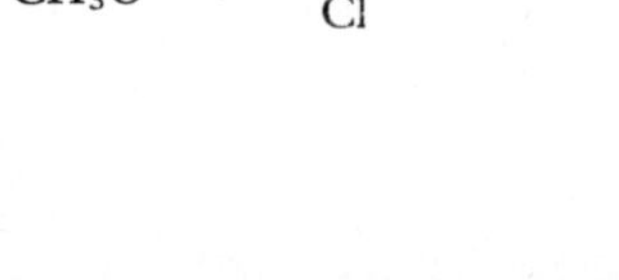

XXXV

Ronnel

which is one of the least toxic organophosphates as far as mammals are concerned.

Several of these newer compounds have such low mammalian toxicities that they are widely used in the control of insects, such as flies and lice, which are intimately associated with farm animals. In a few cases, ronnel being one example, veterinarians apply the organophosphate to the skin of the mammal; it is absorbed and subsequently acts as a systemic, controlling pests feeding on the host tissues. Ronnel is used against the larvae of warble flies, *Hypoderma* spp., which, after hatching from eggs laid on the flanks of cattle, bore through the skin, migrate within the host's body and form large blister-like swellings or warbles on the back, from which they eventually emerge when ready to pupate. Apart from the discomfort caused to cattle, the quality of their hides is greatly reduced by the boring activities of the mature larvae. When ronnel is applied to cattle the young warble larvae are killed soon after they have penetrated beneath the skin.

The possibility of administering organophosphates to farm animals has promoted research into their value for controlling internal helminth parasites as well as insects. In the case of any chemical which is to be administered to an animal an important consideration is the therapeutic index, which is the ratio of the dose which will kill the animal, to the dose which effects a cure. A very high therapeutic index therefore means that there is a big safety margin in the use of the chemical. The first attempts to control internal helminths with organophosphates were made with some of the available insecticides such as ronnel. Promising anthelmintic activity has been shown against parasites of sheep and cattle but the therapeutic index is generally rather low. However, organophosphates are being developed specifically as anthelmintics, examples being the naphthalene derivative known commercially as rametin, and the substance known as haloxon (XXXVI). In the molecule of haloxon the chlorine atoms have been introduced into the two ethoxy-groups.

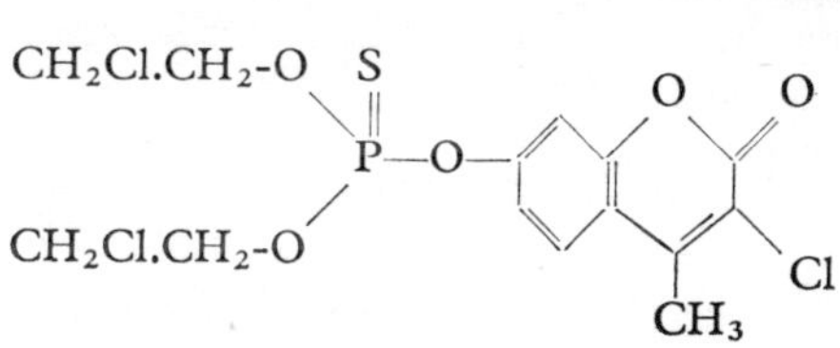

XXXVI

Haloxon

The therapeutic index of each of these compounds is comparatively high, and they have been successfully administered on a large scale for the removal of stomach and intestinal worms from sheep. Formulations of value in cattle worm control are also now being marketed commercially.

The mode of action of organophosphate poisons has received a good deal of attention, and one of their toxic properties is their ability to block the activity of (i.e., inhibit) the enzyme cholinesterase. This enzyme is widely distributed in the animal kingdom and its function is to break down the substance acetylcholine, which plays an important role in nervous activity. In simple terms the passage of nerve impulses along a nerve involves the transmission of waves of electrical activity, and if, for example, this is to bring about the contraction of a vertebrate voluntary muscle (e.g., in a limb), this electrical activity of the nerve must be transformed into mechanical movement. The nerve terminates in an end plate within the muscle, and the arrival of impulses at the end plate causes the release of acetylcholine which diffuses out and stimulates the muscle to contract. Obviously the acetylcholine must then be destroyed, otherwise the muscle would continue to be stimulated, even in the absence of further impulses. Cholinesterase therefore breaks down the acetylcholine as soon as it has performed its function of 'mediator' of the nerve impulse and thus, one of the symptoms of organophosphate inhibition of this enzyme is uncontrolled muscular activity in the form of tremors and convulsions. Furthermore, in vertebrates it is not only the transmission between nerve end plate and voluntary muscle which is 'cholinergic', i.e., requires the mediation of acetylcholine. Many of the functions of the body are not under conscious control (gut movements, heart beat, glandular secretion, etc.) and these involuntary responses are stimulated by the autonomic nerves. These are grouped into two mainly antagonistic systems, the sympathetic and parasympathetic systems. Thus, sympathetic stimulation for example increases the rate of heart beat, whereas parasympathetic stimulation decreases it. The autonomic nerve supply which links the central nervous system with the involuntary muscles or glands (i.e. the effectors) is not completely continuous. At some point there is a gap, or synapse, between the end of the nerve running out from the central nervous system, and the beginning of the nerve supplying the effector. As in the case of the neuromuscular junction, this autonomic synapse is also temporarily bridged when impulses arriving at the synapse cause the release of acetylcholine, which then stimulates the nerve supplying the effector. In the case of the parasympathetic supply, the junction between nerve and effector is also cholinergic, but the sympathetic nerve-effector junction is usually bridged by other chemicals, either adrenaline or nor-adrenaline. Thus many of the symptoms of organophosphate poisoning are those one would expect to result from uncontrolled activity of the autonomic, particularly the parasympathetic, nervous system, i.e., constricted pupils, defaecation, urination, lacrimation, which accompany the twitching and convulsions resulting from the neuromuscular dislocations mentioned above. Death is usually ultimately due to asphyxiation resulting from the malfunctioning of the respiratory muscles of the thorax.

In the case of insects, the hypothesis usually adopted at present is that cholinesterase inhibition is also the primary cause of organophosphate poisoning. Acetylcholine is a prominent feature of the biochemical constition of insect nervous tissue and, once again, the symptoms of poisoning are those one would expect to result from dislocations in the nervous system. Furthermore, there is evidence that synaptic transmission in insects is cholinergic, although the biochemistry of insect nerve tissue has not, as yet, been studied nearly as intensively as that of mammals.

It is interesting that many potent organophosphates are poor inhibitors of cholinesterase *in vitro*, and they are only rendered poisonous as a result of enzyme reactions taking place within the body of the victim. Parathion for example is enzymically oxidised to paraoxon, which is a powerful cholinesterase inhibitor.

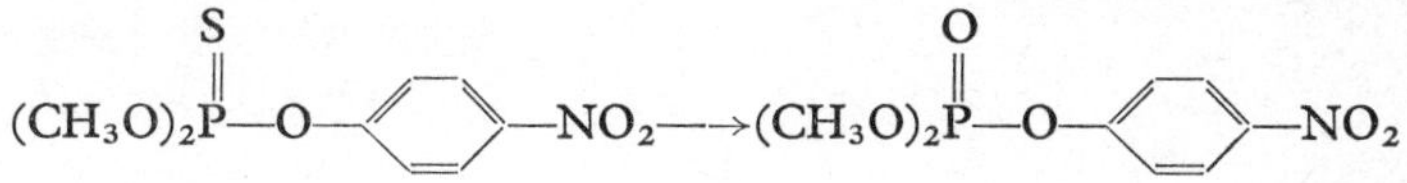

It is not impossible that selective chemicals could be developed which would be 'activated' in the above manner only by enzymes within the body of the target pest, and which would be harmless to other members of the animal community. In fact organophosphates are, in many cases, selective pesticides, some being particularly toxic to vertebrates, some to caterpillars, others to Diptera. A few are useful ovicides, and others primarily acaricides, this selectivity obviously being an improvement on the 'blanket kill' obtained with some of the organochlorine compounds. However, organophosphate selectivity is not primarily due to differences in the activation of these compounds in different species, the more important factors being variations in the metabolic breakdown of the poisons, and their transport, storage and excretion within the animals. Assuming that the primary target in the insect is the nervous system, there are probably differences between species in the chemical and physical barriers which protect this target, but these factors have, as yet, scarcely been investigated.

4.2.6 Carbamates

One outcome of the success of the organophosphates was the development of interest in other compounds with known inhibitory properties towards cholinesterase. This led the chemical industry to the group of substances now known as the organic carbamates, many of which show, to a high degree, the property of selectivity discussed above. Apart from their similar modes of action, other parallels can be drawn between the organophosphates and the carbamates. Just as the organophosphates were considered as derivatives of phosphoric acid, so the carbamates can be related to carbamic acid,

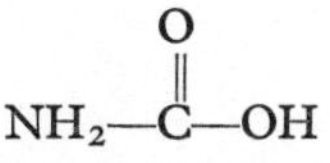

XXXVII
Carbamic acid

a hypothetical acid which exists only in the form of its salts such as ammonium carbamate, NH_2COONH_4. In the case of the organophosphates, the hydrogen atoms of the hydroxyl groups labelled (1) and (2) in formula XXIII are usually replaced by methyl or ethyl groups, and, in a similar manner, one or both of the hydrogen atoms of the carbamic acid $-NH_2$ group will be replaced by a methyl radical in the carbamate pesticides. These structural parallels can be carried further; it was seen that the great variety of organophosphates is the result of the wide range of structures which can be coupled to the oxygen atom labelled (4) in formula XXIII; in just the same way, a considerable variety of chemical groupings can be linked through the oxygen atom of the hydroxyl group in formula XXXVII.

The first insecticidal carbamates, which appeared in the early 1950s, were dimethyl carbamates; i.e., having the $-NH_2$ group replaced by $-N(CH_3)_2$, an example being isolan.

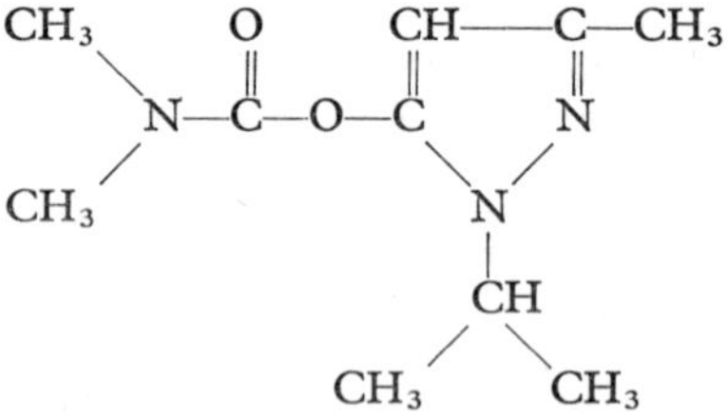

XXXVIII

Isolan

Compounds of this type mainly suffer the same disadvantage as the early organophosphates, i.e., appreciable or high mammalian toxicity (see Table 10). They are generally highly selective, isolan acting as a contact, fumigant and stomach poison to houseflies, and also being highly toxic to aphids, against which it shows systemic properties in the plant host. In some of these compounds selectivity is carried to the point at which it becomes a commercial disadvantage. This may sound curious in view of the desirability of selectivity pointed out previously, but a restricted group of pests has to be of enormous economic or social importance to justify the vast expenditure required to develop a pesticide, highly specific to that group alone (see also Chapter 9). It is necessary to compromise between the indiscriminate poison on the one hand, and the commercially non-viable, but highly specific pesticide on the other. This compromise is achieved by many of the monomethyl carbamates, the first successful member of which was introduced in 1956 under the name sevin.

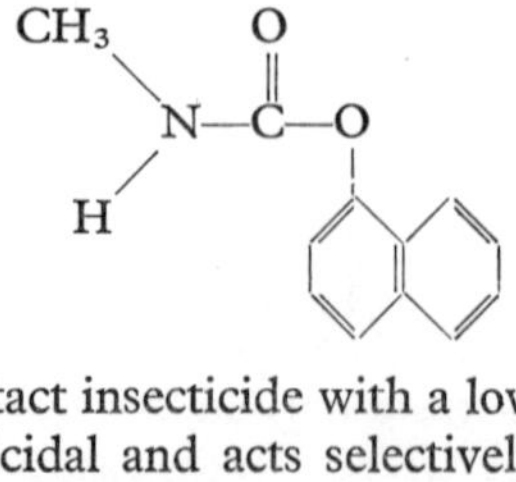

XXXIX

Sevin

Sevin is primarily a contact insecticide with a low mammalian toxicity (see Table 10). It is non-acaricidal and acts selectively on insects. It has been widely used against caterpillar and beetle larvae, particularly bollworms and boll weevil on cotton, and leaf feeding pests of fruit. It has been found particularly useful in the control of codling moth larvae, which burrow into apples, but some common insects, e.g., the adult housefly, are very little affected by contact with sevin. Since this first success with the monomethyl carbamates, the subsequent development of the group has paralleled that of the organophosphates. By coupling various organic groupings to the carbamate hydroxyl group a great variety of compounds have been produced and screened for pesticidal activity. The list of promising compounds is growing steadily, and a multiplicity of examples would serve no useful purpose here. At the time of writing, the commercial marketing of carbamates is on a very much smaller scale than that of organophosphates, but this is undoubtedly one of the active growing points in the present-day chemical control technology.

4.2.7 *Organic amines, amides and their derivatives*

In addition to the carbamates there are other groups of synthetic organic nitrogen compounds which have been found to be valuable pesticides. Prominent among these are various organic amines, i.e., compounds of the type R–NH_2, where R is an alkyl or aryl group. The hydrogen atoms may be replaced by additional organic groups and several long-chain amines, diamines and their derivatives have been tested for the control of *Entamoeba histolytica*, the causative agent of amoebic dysentery in man, and for the destruction of schistosome flukes.

In recent years, other compounds under this general heading have been studied as possible chemosterilants for use in insect control (section 6.2). One example is the triamide derivative of phosphoric acid (XL), which is

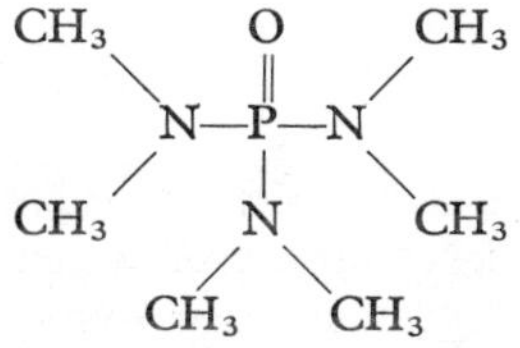

XL

Hempa

known as HMPA or hempa. Hempa is a male sterilant for house-flies and has similar effects on other insect species as well. However, far greater interest has centred upon compounds in which the dimethylamino group is replaced by the aziridine group (XLI).

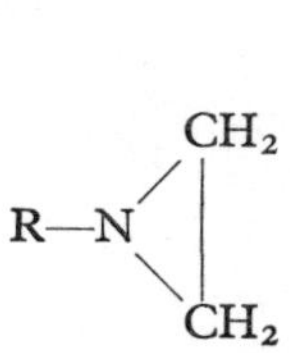

XLI

General formula of Aziridine compounds

These are very active compounds belonging to a class known as the alkylating agents, because of the ease with which they can introduce an alkyl radical into any of a wide variety of molecules with which they react. Many of the most widely studied chemosterilants of the present time such as TEPA (or aphoxide), metepa, tretamine, and apholate belong to the aziridine group. The formula of TEPA (XLII) closely resembles that of hempa (XL).

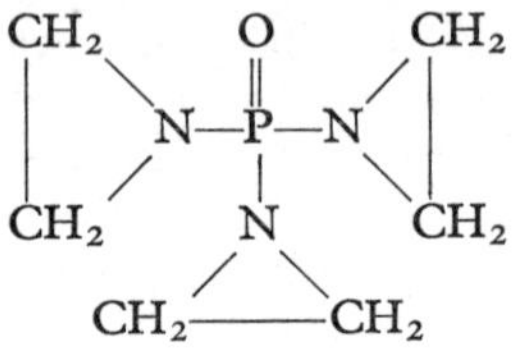

XLII

TEPA

One of the main problems limiting the exploitation of these, and other alkylating agents, for the sterilisation of insect pests on a field scale is the high mammalian toxicity of many of them. Several of them are absorbed through the mammalian skin, and this type of hazard must be circumvented if the transition from laboratory research to the widespread commercial application of chemosterilants is to be successfully achieved.

So far we have considered only alkyl amino compounds, but there are several pesticides in which aromatic groups are coupled to the nitrogen atom. A simple example of a substituted aryl amine is diphenylamine,

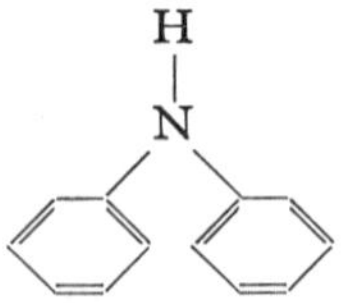

XLIII

Diphenylamine

which is a useful insecticide and acaricide that has been used to kill lice, and to prevent myiasis (the development of dipterous maggots in wounds). Various derivatives of diphenylamine have found some favour as insecticides, and one of them, phenothiazine, has established for itself a prominent position in the history of gut-nematode control.

Phenothiazine can be thought of as diphenylamine with an additional 'bridge' between the phenyl groups formed by a sulphur atom.

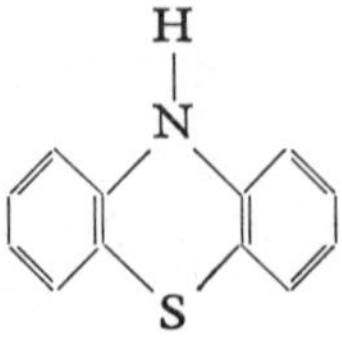

XLIV

Phenothiazine

This substance was first used in the 1930s as an insecticide; it is an effective larvicide for mosquitoes, but it is as an anthelmintic that it has been most widely used. It was first administered in 1938 to control intestinal nematodes in pigs, with only qualified success, but since that time it has been very successfully applied against various nematodes of ruminants such as *Haemonchus contortus*, *Ostertagia* spp., *Trichostrongylus* spp., *Oesophagostomum* spp., and *Chabertia ovina*. It has also been used to control *Strongylus* spp. in horses, and *Heterakis gallinae* in chickens. In many cases, even more effective compounds are now available for dealing with these parasites and the use of phenothiazine is declining. It still retains some value however, and substantial amounts are still being manufactured. There is, in fact, good evidence of synergistic effects (section 3.2.2(b)) between phenothiazine and other widely used anthelmintics, such as dichlorophen (X), and various organophosphates.

Phenothiazine has a very low solubility in water and is usually administered to animals either as a liquid suspension, in tablet form, or mixed with the food. A single large dose is the normal treatment for a heavily parasitised animal, but in sheep exposed continuously to parasitic reinfestation, smaller doses administered monthly may afford prolonged protection. On occasions, sheep have been treated with a small dose each day, with the result that the egg laying capacity of female nematodes in the gut is depressed. The lowered number of eggs are passed out in the host faeces together with phenothiazine which kills off the young larvae as they hatch.

One of the disadvantages of phenothiazine is its indifferent action against certain species of roundworms. Efforts to increase its effectiveness in these cases have included the marketing of finely ground or 'micronised' phenothiazine, since it is known that the success of the treatment is inversely proportional to the size of phenothiazine particles administered to the host animal. Phenothiazine is a contact pesticide and the reduction in particle size increases the efficiency with which it is absorbed through the cuticle of the worm. The chemical is also absorbed by the gut of the host and the fact that it is less toxic to the host than to the worm appears to depend upon the ability of the mammal to break it down readily to harmless metabolites, whereas it accumulates unchanged in the parasite. The parasite is not necessarily killed by phenothiazine but it may be affected in such a way that the host can eliminate it live from the body or, if it remains, its egg laying ability is greatly depressed as mentioned above. The biochemical explanation of these effects has so far eluded researchers, but there is some suggestion of anticholinesterase properties in this chemical.

The variable effects of phenothiazine on different helminth species and its considerable toxicity to some host species (e.g. carnivores) are disadvantages which have led to its gradual supersedure by other pesticides. To these disadvantages must be added the hypersensitivity to light which may occur in treated cattle, the red colour which may be imparted to the milk, and the development of resistance which has occurred in some strains of *Haemonchus contortus* in sheep and strongyle worms in horses. A general discussion of the pesticide resistance problem is given in section 8.2.1.

The phenothiazine group of chemicals are not the only organic nitrogen compounds which provide valuable anthelmintics, and brief mention must be made of two other groups, the quaternary ammonium compounds and the heterocyclic nitrogen compounds.

4.2.8 Quaternary ammonium compounds

These may be considered as derivatives of the ammonium ion, which can be illustrated by reference to a simple substance like ammonium hydroxide $NH_4^+OH^-$. The hydrogen atoms of the cation can be replaced by organic groups and it can be associated with anions other than the hydroxyl ion, thus forming complex organic salts. An example is bephenium hydroxynaphthoate.

XLV

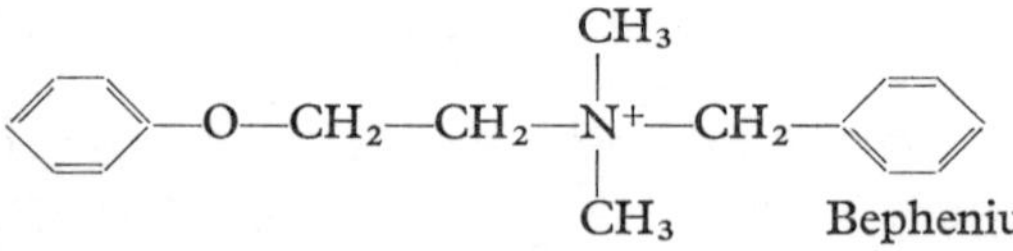

$C_{11}H_7O_3^-$

Bephenium hydroxynaphthoate

This anthelmintic was introduced in 1958 and has been very successfully used against hookworms and *Ascaris lumbricoides* in human beings, and against the various trichostrongyle species of nematodes in sheep, being especially useful against *Nematodirus* spp. Bephenium salts are very poorly absorbed by the mammalian gut, and the small amounts which pass this barrier are readily metabolised, so that as a result they appear to be 'safe' anthelmintics. The way in which they attack the parasite is somewhat obscure, but the mode of action may be effected via the nervous system since similarities have been noted between their effects and those of acetylcholine.

The pesticidal value of complex quaternary ammonium compounds is not confined to their use as anthelmintics, some have occasionally been employed as insecticides, and in addition a number of the drugs which have been used to attack the trypanosomes causing sleeping sickness in man and cattle belong to this category. The first of these, introduced in 1949, was the substance quinapyramine, or 'antrycide'.

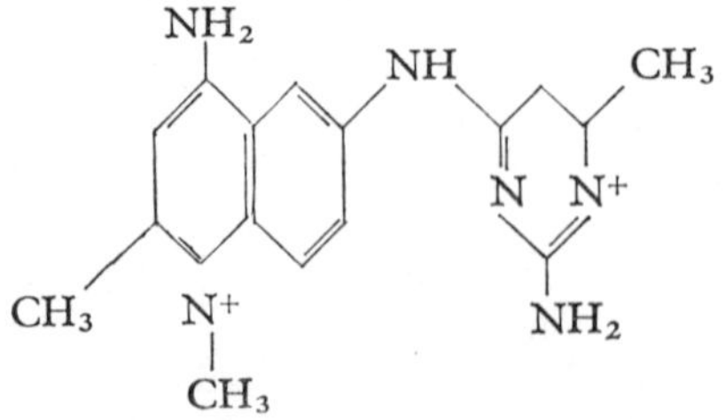

XLVI

Antrycide

This has been widely used on cattle and is a good prophylactic drug, i.e., it affords protection for some time after administration. In the case of antrycide this period of protection appears to last from one month up to three or four months, it being most effective when the density of the tsetse fly vector is low. For some time antrycide was used almost exclusively to treat trypanosomiasis, but in the 1950s other quaternary ammonium derivatives appeared, such as prothidium (XLVII) and ethidium (XLVIII), compounds.

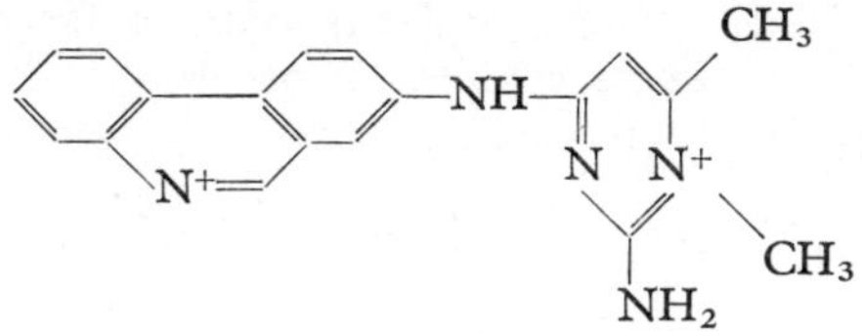

XLVII

Prothidium

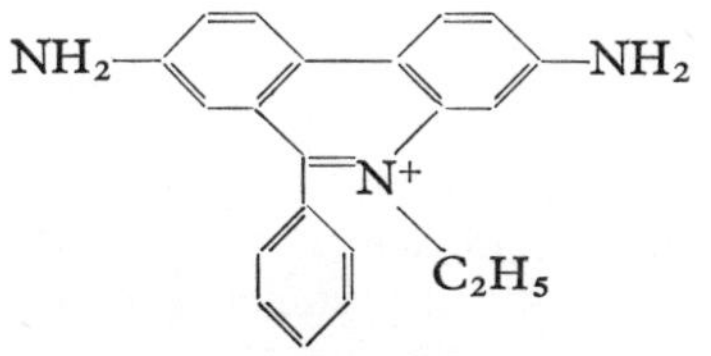

XLVIII

Ethidium

The structural similarities with antrycide are readily recognisable.

The social and economic consequences of human and domestic animal trypanosomiasis are so important that a very great deal of research has been devoted to the discovery of trypanocides. The above examples are representative of a very considerable number of chemicals available today, which cannot all be considered within the confines of this book. The number of chemicals grows each year, partly because many of the existing drugs such as antrycide have evoked the development of resistant strains of trypanosomes. Any prophylactic given regularly over a long period is likely ultimately to produce resistance (section 8.2.1) and the countermeasure adopted is to use curative drugs as well, which rid the animal of trypanosomes but which unfortunately in no way prevent reinfection.

4.2.9 Heterocyclic nitrogen compounds

The inclusion of nitrogen within an organic ring structure is a feature of antrycide (XLVI) and this compound could have been discussed under this heading. It was, however, classified for convenience on the basis of its quaternary ammonium relationships. There are a number of valuable pesticides which can conveniently be considered together on the basis of their possession of a heterocyclic ring. We have already discussed some compounds of this type, the naturally occurring alkaloids (section 3.2.2b), but here we are concerned with synthetic compounds containing nitrogen linked in this manner. Some of these synthetic heterocyclic compounds have obvious relationships with the naturally occurring alkaloids. The success of nicotine as an insecticide was followed by the synthesis and use of closely related compounds such as dipyridyl and neonicotine. The parentage of many of the synthetic antimalarial drugs, such as choroquine, endochin and mepacrin can be traced back to the naturally occurring alkaloid, quinine. In some cases these synthetic compounds have been successfully used on quite new pest problems as well, a good example being the use of mepacrin and related compounds against tapeworms.

There are, however, a considerable number of heterocyclic pesticides with no direct relationship to compounds occurring naturally in plants. A simple starting point for discussing heterocyclic molecules is of course pyridine.

XLIX

Pyridine

Pyridine is of little pesticidal importance, although it has had some minor application as an acaricide. The introduction of a side chain to produce methoxyethyl pyridine gives a valuable pesticide, generally known as methyridine.

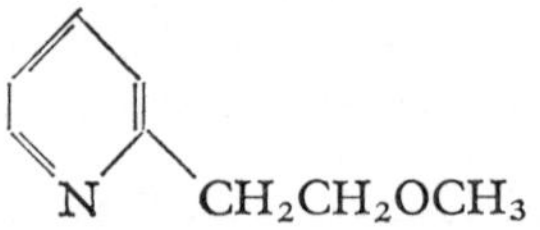

L

Methyridine

Methyridine is a valuable anthelmintic, used against the stomach, intestinal, and lung worms of sheep and cattle. One curious feature of methyridine is that in addition to normal oral dosing, it is equally effective if injected subcutaneously into the host animal. The chemical is apparently excreted into the gut lumen of the host where it brings about the elimination of the worms located there. It is one of several really effective nematicides that can be used with relative safety on cattle. There are indications that it may also successfully deal with another difficult parasitological problem, the removal of *Capillaria* spp. from the crops, intestines and caeca of various domestic fowls, to which it may be administered either as an injection, or in the drinking water. Methyridine causes nematode muscle to contract and become paralysed and there is some eividence that it acts on the neuromuscular synapse, probably in a similar manner to acetylcholine.

Heterocyclic compounds may have more than one nitrogen atom incorporated into the ring, as for example in the substance piperazine.

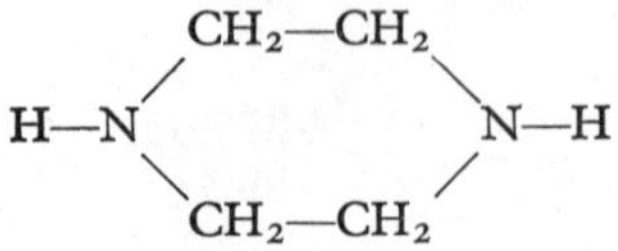

LI

Piperazine

This chemical has no insecticidal properties but its valuable anthelmintic activity was recognised in 1949. Its low toxicity and high level of effectiveness have led to its wide use against human helminths, such as ascarids and the pinworm *Enterobius vermicularis.* The common ascarids of dogs, cats, pigs, and chickens also yield to treatment with piperazine salts, and their use against horse ascarids also provides a bonus in the sense that a proportion of the strongyle worms are destroyed as well.

The site of action of piperazine again seems to be the neuromuscular junction and it produces a flaccid paralysis in ascarids which prevents the contraction which one can normally induce with acetylcholine. The paralysed worms are passed out by the host, the vertebrate neuromuscular system being, for some as yet unknown reason, largely unaffected by the therapeutic dose of piperazine.

In addition to salts of piperazine, a number of its derivatives have also been used as anthelmintics. Probably the most successful pesticide that can be classified under this heading is diethylcarbamazine, or hetrazan.

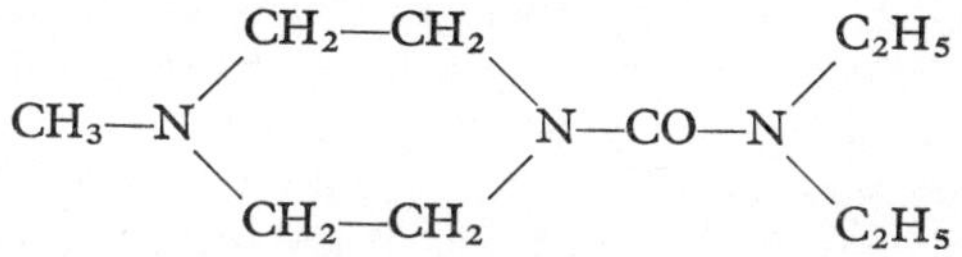

LII

Hetrazan

This was used as early as 1948 against ascarids, but piperazine was cheaper and more effective. However, in the late 1950s it was found that diethylcarbamazine was effective in reducing the number of lungworms causing 'husk' in cattle. The adult lungworms, *Dictyocaulus viviparus* lives in the bronchial passages and its larvae cause considerable damage by burrowing in the lung tissue itself. Intramuscular injections of diethylcarbamazine affect both adult and larvae, the latter being particularly susceptible. The adult numbers are reduced but the host animal is not necessarily completely cured.

The other major use of diethylcarbamazine is in the treatment of human patients infested with filarial worms. These tiny nematodes may cause elephantiasis (*Wuchereria bancrofti* and *Brugia malayi*), loaiasis (*Loa loa*) or onchocerciasis (*Onchocerca volvulus*). Vast numbers of microfilariae (the juvenile stage) swarm in the blood or subcutaneous tissues of the patient, some are picked up by the biting flies in which they undergo development, and are subsequently transmitted when these insects bite again. Control involves the use of insecticides on the vectors, and the use of orally administered diethylcarbamazine, which renders the microfilariae susceptible to phagocytosis by the liver. Shortly after administering the drug the microfilariae disappear from the blood or skin, thus removing the infective risk presented by the patient. However, the use of the drug causes such a sudden destruction of the worms that the release of their proteins may result in allergic and other reactions in the patient. This is particularly true in cases of onchocerciasis, and the treatment requires considerable skill.

Space does not permit discussion of many of the more elaborate heterocyclic structures which have been used from time to time in pest control. However, there is one compound, in which three nitrogen atoms are incorporated into ring structures, which is so important at the present time that it merits some consideration. This compound is known as thiabendazole, and has the following structure.

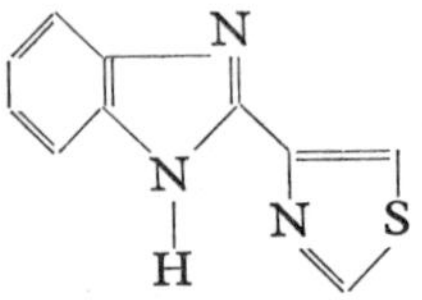

LIII

Thiabendazole

Investigations into the activity of thiabendazole were first published in 1961, when it was shown to be an effective anthelmintic for both immature and adult roundworms in a wide range of hosts. In sheep the chemical removes species belonging to the genera *Trichostrongylus*, *Cooperia*, *Nematodirus*, *Ostertagia*, *Haemonchus*, *Oesophagostomum*, *Bunostomum*, *Strongylus* and *Chabertia*, and the therapeutic index is high. Similar successes have been reported against hookworms, whipworms and other roundworms in dogs, and against ascarids and strongyles in horses. At the present time, thiabendazole is at least as effective as the commonly used bephenium compounds and it figures very prominently in the commercial market, although there have been recent reports of resistant strains of *Haemonchus contortus* appearing.

A number of promising new pesticides can be loosely classified under the heading of heterocyclic compounds. An example is N-tritylmorpholine (LIV), a valuable molluscicide which is used to destroy the secondary hosts of various helminth parasites.

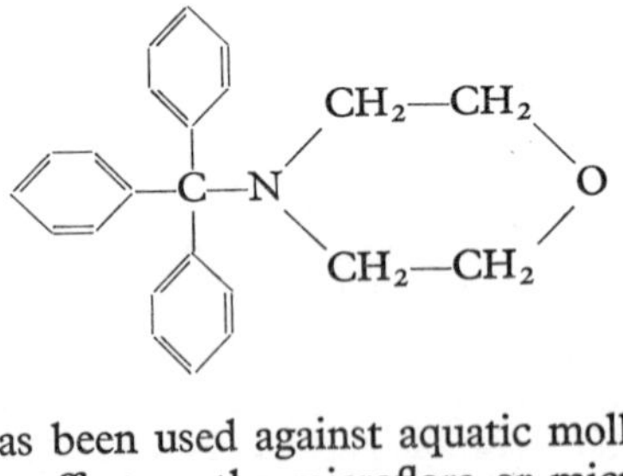

LIV

N-Tritylmorpholine

This compound has been used against aquatic molluscs on a large scale, without any apparent effect on the microflora or microfauna of the treated water. It is also non-phytotoxic, but is lethal to snails, even at high dilutions.

More direct control over the snail-borne diseases, referred to as schistosomiasis, is achieved by another comparatively new pesticide, ambilhar which is effective against the adult worms.

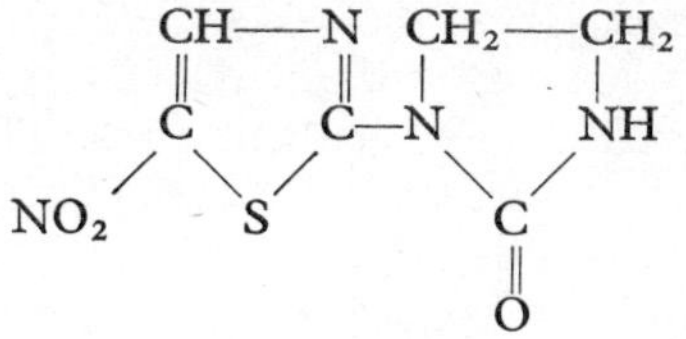

LV

Ambilhar

Oral administration, or intravenous injection of ambilhar are effective in destroying the blood worms responsible for the various forms of schistosomiasis. The therapeutic index of this compound is high which is one of its advantages over the older antimony compounds (section 4.2.1) which are still quite widely used.

Another compound which, like thiabendazole and ambilhar, contains sulphur as well as nitrogen within its ring structure is tetramisole.

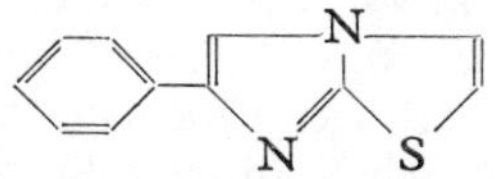

LVI

Tetramisole

This was originally synthesised as a mixture of the dextro and laevo isomers and the activity lies in the L component. This anthelmintic has been shown to be very effective in eliminating the gastro-intestinal roundworms of sheep and the lungworms of sheep and pigs. It is administered orally or by injection, and again shows a high therapeutic index.

The final example of these rather complex pesticides which contain heterocyclic rings is the substance norbormide, which rejoices in the chemical name of 5-(α-hydroxy-d-2-pyridyl-benzyl)-7-(α2 pyridylbenzylidene) norborn-5-ene-2, 3 dicarboximide.

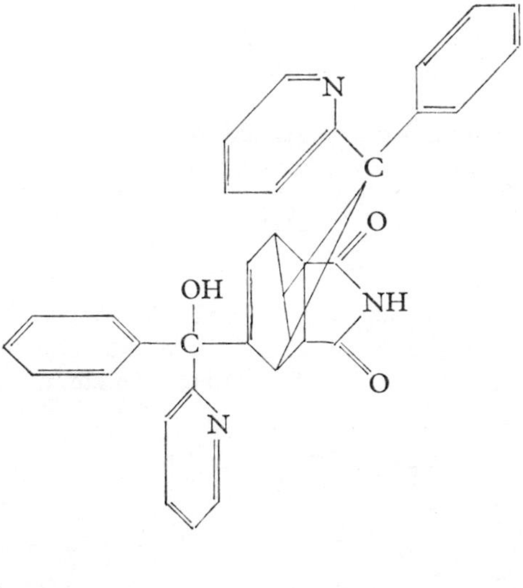

LVII

Norbormide

Norbormide is a rodenticide, introduced in the early 1960s, which is highly specific, being extremely toxic when ingested by rats, but virtually harmless to domestic and farm animals, mice, fish and birds. The poisoned rat shows characteristically a loss of co-ordination in the hind legs, laboured breathing and blanching of the extremities. This blanching appears to be due to the violent constriction of the blood vessels which also produces a local anaemia in the internal organs including the heart. The animal appears to die from shock, but the precise mode of action of this chemical is as yet unknown. It has proved of value in dealing with rats which have developed resistance to warfarin, the most commonly used rodenticide at the present time (section 4.2.11 and section 8.2.1).

4.2.10 Organic sulphur compounds

The use of inorganic sulphur compounds in pest control has been discussed, but this element is also widely distributed in organic pesticides. The simplest example that can be quoted is that of carbon disulphide, CS_2, which is a heavy yellow volatile liquid and has been used for over a hundred years as a fumigant. It has proved useful when injected into the soil against pests, and is used to fumigate stored products such as potatoes and grain. It is also incorporated into fumigant cartridges for the destruction of moles. The extreme inflammability and toxicity of carbon disulphide are serious disadvantages which mean that great caution is necessary when it is used, and the tendency now is to adopt safer alternatives whenever possible.

More complex pesticides, incorporating sulphur into an organic molecule, have already been mentioned in this chapter, for example in the case of the 'thio' group of organophosphates. However, we have not yet considered a number of valuable pesticides, all of which may be regarded as derivatives of benzene sulphonic acid.

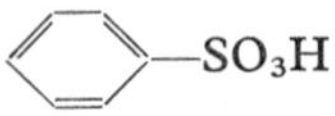

LVIII

Benzene sulphonic acid

One very famous group of drugs is related to the amide derivative of benzene sulphonic acid, and is therefore known as the sulphonamides.

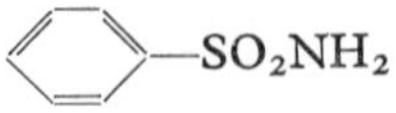

LIX

Sulphonamide

In the 1930s it was found that streptococcal infections could be cured by the injection of a drug called prontosil, and it was subsequently found that the activity of prontosil was due to the release of a sulphonamide, known now as sulphanilamide.

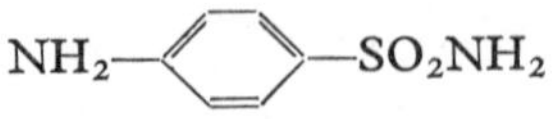

LX

Sulphanilamide

The immediate result of this discovery was the synthesis of many derivatives of sulphanilamide, a number of which proved to be invaluable drugs for use against bacterial infections. It was then discovered that in addition to their bacteriacidal action, some of the sulphonamides also affected parasitic protozoa in various animals. They have been used in the treatment of malaria, of toxoplasmosis, and to a limited extent in the treatment of infections caused by parasitic amoebae. However, their major value in the control of parasitic protozoa has been their use against the sporozoans causing the disease known as coccidiosis. The coccidial organisms are primarily intracellular parasites of the gut epithelium, causing haemorrhage and death in a variety of animals including poultry, cattle, sheep and rabbits.

As early as 1939 it was found sulphanilamide (LX) produced a beneficial effect in chickens suffering from coccidiosis, but several of the derivatives of sulphanilamide subsequently proved of greater value. The first to be used was sulphaguanidine

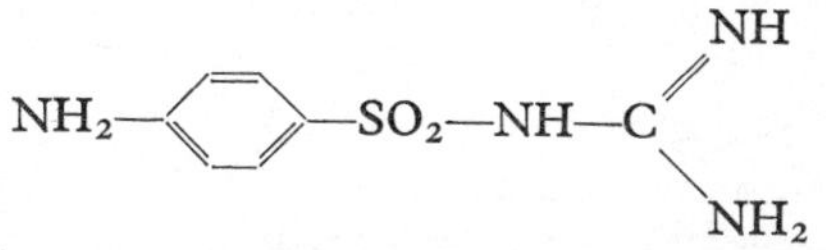

LXI

Sulphaguanidine

which was administered by mixing with the feed of infected animals. However, in chickens, the effect of sulphaguanidine appears to be mainly prophylactic rather than curative, and although this chemical is still used in the treatment of other animals, birds have responded better to the use of compounds such as sulphamethazine (LXII) and particularly sulphaquinoxaline (LXIII).

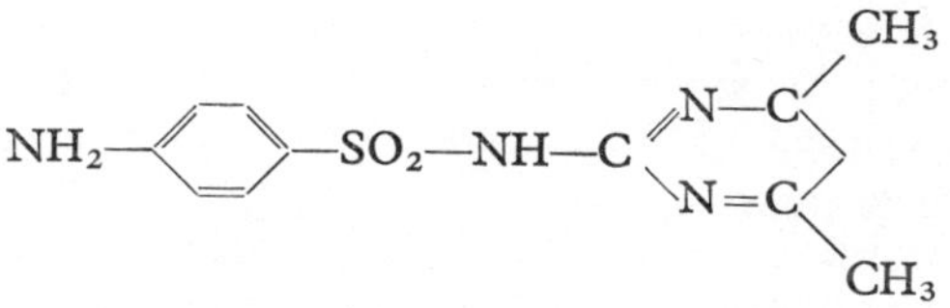

LXII

Sulphamethazine

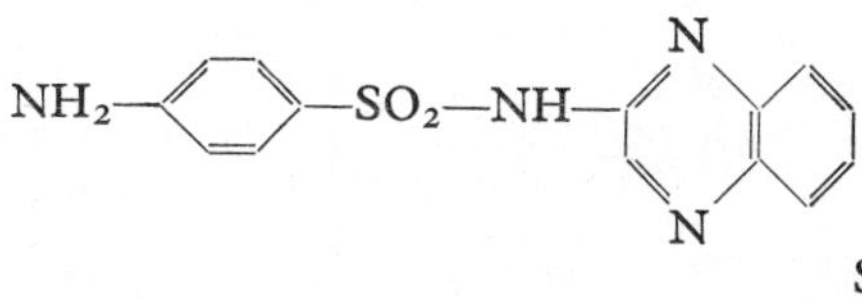

LXIII

Sulphaquinoxaline

The success of the sulphonamides as anti-coccidial agents is accompanied by the disadvantage of their substantial toxic effects on the host. Furthermore, the problem which invariably accompanies the prophylactic administration of a drug, or indeed the prolonged application of an insecticide, has arisen on several occasions, namely the breakdown in control as a result of the development of resistance.

The amides are not the only derivatives of benzene sulphonic acid in use as pesticides. If we replace the $-NH_2$ group by a phenyl group we have the basis for a series of 'sulphones', some of which are widely used as acaricides. The simplest example is diphenyl sulphone (LXIV) which has an ovicidal action against the summer eggs of the fruit tree red spider mite.

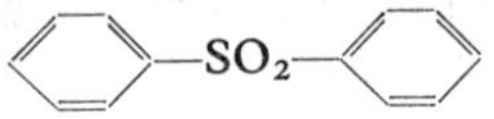

LXIV

Diphenyl sulphone

The introduction of a chlorine atom gives 'sulphenone' (LXV) which is acaricidal to various red spider mites and which has hardly any insecticidal effect on insect parasites, predators or honey bees.

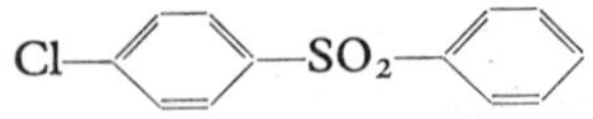

LXV

Sulphenone

If the number of chlorine atoms is increased to four, we have the very valuable acaricide known as tedion, or tetradifon.

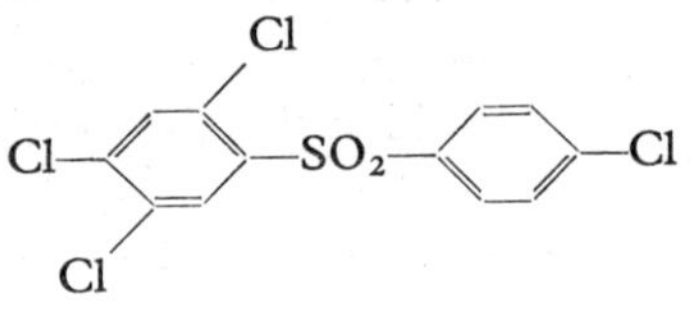

LXVI

Tedion

Tedion affects all stages of the red spider mite life cycle, although in the case of the eggs it appears merely to delay hatching rather than kill them. It is a very persistent contact poison which also shows some systemic properties in its ability to pass through from one side of the leaf to the other. It has been quite widely used since its introduction in 1954, its advantages including very low toxicity to beneficial insects and to plants, and an almost complete lack of toxicity to warm-blooded animals (see Table 10).

If the hydrogen atom of the benzene sulphonic acid radical is replaced by a chloro-phenyl group, we have a second series of derivatives, the phenyl benzene sulphonates. Of these, the most effective acaricide is the highly persistent substance introduced in 1949, and known as chlorfenson, or ovex.

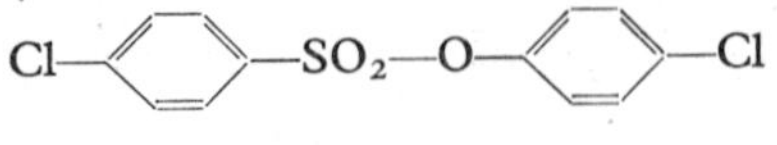

LXVII

Chlorfenson

This acaricide has little effect on the adult mite, but is used primarily against the newly hatched nymph and as an ovicide. It again shows a pronounced lack

PLATE 4A SCHISTOSOMIASIS: CONDITIONS FOR PERPETUATION
Both the washing of clothes and bathing in contaminated water aids the perpetuation of Schistosomiasis (see also Plate 4B). Note the swollen abdomen of the boy to the left of the picture; this is one of the clinical signs of Schistosomiasis.

PLATE 4B SCHISTOSOMIASIS: CONDITIONS FOR PERPETUATION OF SCHISTOSOMA JAPONICUM

In cultivating the paddy field for rice growing, the farmer wades in water and hence comes into contact with the stage of the Schistosome life cycle (miracidium) which penetrates human skin and hence develops into its blood dwelling habit.
(*Courtesy The World Health Organisation*)

of toxicity to other animal groups and shares with tedion the pseudo-systemic properties mentioned above.

It should be pointed out that the acaricidal properties of these compounds do not depend on the presence of the sulphonate or sulphone group, but rather on the possession of the two phenyl groups linked by some intermediate bridge. This bridge may be formed in other ways, as for example in the useful acaricide, chlorbenside.

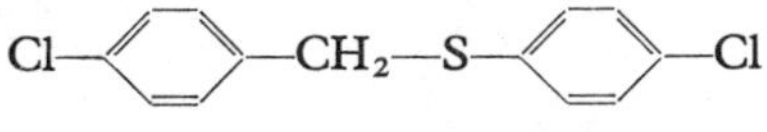

LXVIII
Chlorbenside

In fact, although these diphenyl compounds have been classified for convenience under the heading of organic sulphur compounds, it is possible to obtain acaricides of the diphenyl type in which the bridge group does not contain sulphur at all, one example being the glasshouse acaricide which was among the first of this type to be introduced in 1945, the substance azobenzene.

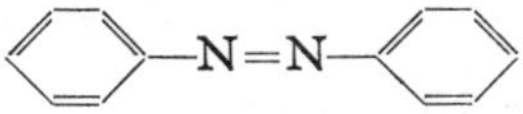

LXIX
Azobenzene

Indeed, structural parallels can be drawn between the arrangement of the benzene rings in these compounds, and that in the DDT group of chlorinated hydrocarbons. However, little insecticidal activity is found in any of these acaricides, and their precise mode of action on the mites is not completely understood.

Before leaving the organic sulphur compounds brief mention must be made of the thiocyanates (LXX) and their structural isomers the isothiocyanates (LXXI).

$R{-}S{-}C{\equiv}N$ LXX Thiocyanates

$R{-}N{=}C{=}S$ LXXI Isothiocyanates

When R is an alkyl group a number of very useful fumigant compounds are obtained. Because of plant toxicity they have been mainly employed in soil injection projects, methyl isothiocyanate for example being widely used against eelworms. The allyl derivative has been used against grain beetles and soil pests such as chafers and click beetles. More complex organic thiocyanates, known as lethanes, were used as contact insecticides during the 1939–45 war, but they have now been almost completely superseded by other compounds.

Other organic compounds containing nitrogen and sulphur are employed as pesticides. Thiourea (LXXII) is poisonous to dipteran larvae and its α-naphthyl derivative (LXXIII), which is known by its initial letters as ANTU, is a widely used rodenticide.

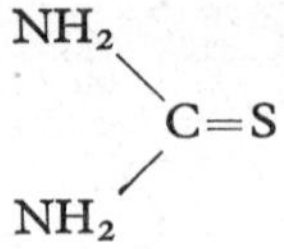

LXXII

Thiourea

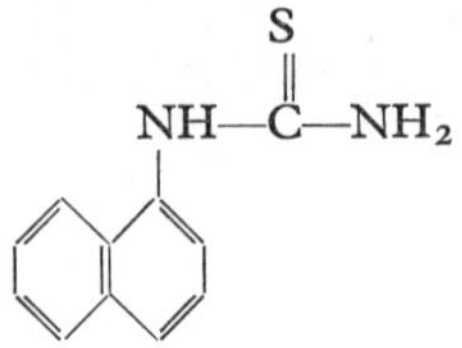

LXXIII

Antu

ANTU was discovered almost by accident during experiments on taste in rats. Some human beings can taste phenylthiourea, others cannot, this being genetically controlled. When this work was extended to rats they were killed, and the most toxic but most palatable derivative was found to be ANTU. It is relatively non-toxic to man and domestic mammals and is only effective against one species of rat, the common, brown or Norway rat, *Rattus norvegicus*.

4.2.11 Miscellaneous compounds

In the preceding pages it has been possible to do little more than indicate, with examples, some of the main chemical classes of compounds to which both modern pesticides, and those of the past, belong. The total number of chemicals which have been used in pest control at some time or other runs into thousands and this account has, of necessity, been extremely superficial. Some very well known compounds do not fit conveniently under any of the subheadings used in this chapter, often because only one member of a chemical series finds widespread commercial use as a pesticide.

A good example is the simple gaseous substance hydrogen cyanide (HCN), which is a powerful respiratory poison used as a fumigant for such problems as the destruction of insect pests in stored products, and for the gassing of rabbits. Hydrogen cyanide is lethal to animal life generally and the dangers associated with its use are such that special legislation has been enacted in Great Britain specifically to ensure that fumigation is carried out under strictly controlled conditions. In a sealed food store, the gas can be generated by the action of acid on a cyanide salt, but the more usual procedure these days is to employ cylinders of compressed gas.

Another example of a widely used compound which can be included under this heading is metaldehyde, which is a polymer of acetaldehyde (CH_3CHO) and is used on a very large scale as a molluscicide. The importance of slugs and snails is very considerable in the context of the problems discussed in this book. The damage which they do to crops is very substantial in the moist regions of a country, and snails have already been described as important intermediate hosts in the life cycles of the organisms causing such parasitic diseases as schistosomiasis in man, and fascioliasis in sheep and cattle. Metaldehyde is usually administered to molluscs in a bait, the result of ingestion being the copious production of mucus, paralysis and ultimately death.

Finally, a pesticide which has been mentioned at several points in this book in connection with rodenticide resistance, and which must receive some attention in this chapter, is the substance warfarin. This is a coumarin, i.e., a compound containing the nucleus shown in LXXIV, which appears in a number of different pesticides (see for example the organophosphate haloxon, XXXVI).

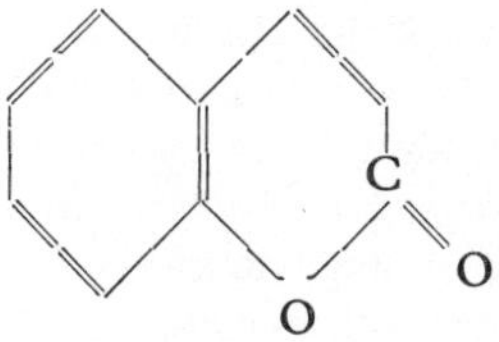

LXXIV

Coumarin group

The origin of warfarin can be traced back to 1941 when it was discovered that one member of this class of compounds, dicoumarin, was responsible for the poisoning of cattle by sweet clover. This discovery was exploited by instituting a search for related compounds which could be used to kill rodents and the most effective was 3(α-acetonyl-benzyl)-4-hydroxycoumarin, also known now as warfarin.

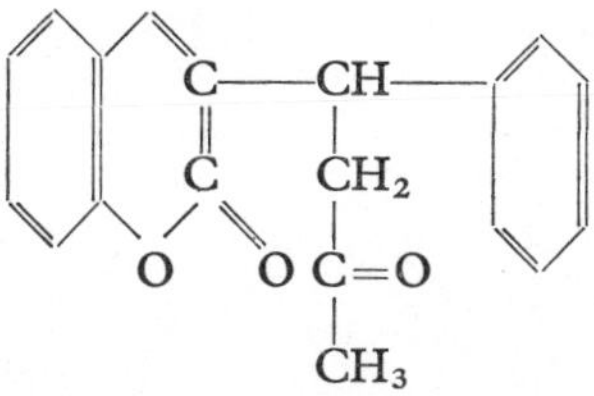

LXXV

Warfarin

The addition of a chlorine atom into the para position of the right hand benzene ring gives another useful rat poison, coumachlor. Warfarin is very toxic to rats, a 1:4000 dilution in baits being effective, and mice are similarly affected. However on a unit weight per body weight basis (mg poison/kg body wt) man and domesticated mammals are about as susceptible as rodents, so care has to be exercised in the use of warfarin baits. The coumarin poisons are termed 'chronic' poisons in that the effects are delayed, illness overtaking the rat some time after it has fed. As a consequence it does not associate the illness with its food, as in the case of an acute poison such as zinc phosphide, and it will return to the baits to feed again. A lethal dose is thus gradually built up, one of the effects of the chemical being to interfere with the synthesis of prothrombin which is essential in the normal blood clotting process.

Affected rats therefore frequently bleed to death as a result of minor haemorrhages, and this type of poison is consequently known as an anti-coagulant poison. Vitamin K, which promotes the production of prothrombin may be used as an antidote in the case of accidental poisoning in domestic animals.

In spite of the numerous control measures which are reviewed in this book, many present-day pest problems respond only to the use of chemicals, referred to elsewhere as 'the expensive flyswatter' technique. The basic failure to control pests, by permanent increases in the environmental pressures acting on them, means that the commercial rewards for the development of effective pesticides are considerable. In consequence, a large number of new compounds appear on the market each year. This chapter is intended to provide a 'chemical framework' on the basis of which many of these new chemicals can be classified, by which their mode of entry and action may be understood, and by which they may be placed in context with many of the notable discoveries of the past. However, it must be appreciated that in any such rapidly developing field, new chemicals from quite distinct groupings will continually be appearing, and no claim to provide an exhaustive and up to date survey of this field could ever be justified.

Chapter Five
THE FORMULATION AND APPLICATION OF PESTICIDES

5.1 Formulation

Each year the chemical industry empirically tests, or 'screens', many thousands of new chemicals for pesticidal activity. Very few of these chemicals will ever actually be used as pesticides, the majority being rejected for a multiplicity of reasons which are the outcome of the exacting specifications of the present time. For example, a potentially useful anthelmintic may have too low a therapeutic index, or a potentially useful insecticide may be excessively damaging to plant tissue, or have a dangerously high mammalian toxicity. The few remaining compounds will only realise their pesticidal potential if methods can be devised for efficiently distributing them to the sites of pest activity, and ensuring that they attain an effective concentration at those sites. These properties are frequently obtained by mixing, or 'formulating', the pesticide with other chemicals. The complexity required in the formulation of a pesticide largely depends upon the complexity of the application technique. Obviously crop spraying with insecticides raises rather different problems to mixing rodent baits or dosing a farm animal with an anthelmintic. In the latter case very little formulation may be involved, the pure chemical may be mixed at the correct dose in the food or water and the animal allowed to ingest it, an example being the administration of some of the sulpha drugs to chickens suffering from coccidiosis (section 4.2.10). However, even with this simple type of administration attention may have to be given to the form in which the pesticide is used, the importance of particle size, for example, being well illustrated by the micronised phenothiazine preparations described in section 4.2.7. An important factor in the therapeutic treatment of parasite infestations by oral administration is to ensure the release of an adequate concentration of the anthelmintic in the infested part of the gut. This may be possible by a direct method, as for example the use of a stomach tube in the administration of CCl_4, but more usually the anthelmintic is enclosed in a bolus, or in a capsule which will break down as a result of the digestive activity in the required part of the gut. Frequently soft gelatin capsules are used, but if for example the nature of the chemical is such that it will cause irritation in the upper digestive tract, then a harder capsule, or a tablet formulation will be used. Stomach worms in sheep have been treated with a so-called 'explosive bolus' containing phenothiazine which is formulated with an effervescent salt, and which undergoes a vigorous reaction in the acidic stomach juices, releasing the phenothiazine where it is required.

In some cases parasitic disease may be treated by injecting the pesticide into the host, either intravenously, intramuscularly or subcutaneously. The pesticide must be formulated in such a way that the injection causes the minimum of shock or of local irritation to the host animal. Formulation may affect the behaviour of the pesticide in the host, as for example in the case of intramuscular injections, where an aqueous solution is absorbed rapidly, whereas a finely divided suspension in a vegetable oil is released and distributed at a slower rate. The vegetable oil acts simply as a carrier for the pesticide, and it is ultimately hydrolysed by the tissues and eliminated.

The use of carriers in this way is very common in the application of insecticides. In former times the control expert could sometimes apply the pure pesticide to a crop, untroubled by the need for the so-called auxiliary substances of modern pesticide formulations, as for example in the case of the dusting of plants with flowers of sulphur. However, even the now obsolescent use of inorganic poison dusts usually involved the incorporation of some sort of inert diluent or 'filler'; and with modern insecticides the cost, toxicity, and often the phytotoxicity, of the pure chemical is such that considerable dilution is usually essential. In the case of indirect control procedures it is important to place an adequate covering of chemical over the surfaces which will be subject to attack. In the case of direct control procedures it is essential to ensure that as many insects as possible come into contact with a lethal dose of the poison. In both cases the aim is to achieve the maximum possible effect with the minimum possible expenditure on chemicals. Among the factors which ensure efficient application are formulation, and also the design of application equipment (considered in the next section). The first consideration governing insecticidal formulation is the form in which the insecticide is to be applied, and Table 11 lists the various methods commonly used at the present time.

It has already been mentioned that dusting was the original application technique for crop protecting pesticides. Dust formulations still retain certain advantages over liquid formulations in that they are generally less bulky and hence less expensive to apply; and in areas where water shortage is a problem they will obviously be preferred to sprays. One of the major problems in crop dusting is to ensure adequate retention of the very small particles on the leaves. Many particles fail to penetrate the boundary layers of air, and thus do not impact on the leaves, and even when contact does take place, the particles frequently fail to adhere. The scope for modifying dust formulations so as to increase adhesion of particles to leaves is limited, since one may well also increase the adhesion of particles to one another in the dusting machine. Some improvement can be obtained by modifying the application equipment

TABLE 11 VARIOUS FORMS IN WHICH INSECTICIDES CAN BE APPLIED

1. *Solid formulations*
   - (a) Smokes — Particle diameter usually 0·2–2μ
   - (b) Dusts — Particle diameter usually 1–25μ
   - (c) Granules — Particle diameter usually 200–100μ
   - (d) Baits — Insecticide mixed with food or other attractive material
2. *Liquid formulations*
   - (a) Fog formulations—Droplet size 5–30μ
   - (b) Spray and mist formulations
     - (i) True solutions
     - (ii) Suspensions
     - (iii) Emulsions

     } Droplet size very variable (40–400μ range) depending upon machine used.
   - (c) Dips, seed dressings, liquid baits, impregnation and irrigation solutions
3. *Vapour formulations*
   - (a) Gases
   - (b) Highly volatile solids and liquids
4. *Insecticidal surfaces*

   These techniques consist of mixing insecticides in plastic or resin strips, or alternatively in paint, with the result that insecticide is slowly released at the surface.

(see next section), but in general, the proportion of applied insecticide retained by the crop is low for dust formulations, values as low as 20% being quoted by some authors. Nevertheless poor retention can under some circumstances be advantageous. If, for example, insecticidal treatment is required shortly before a crop is to be harvested for human consumption, dusting may be preferable to spraying, because the deposits weather more rapidly and thus tend to have less persistence than a corresponding spray deposit. In general, however, dust formulations are less popular than spray formulations, one of the reasons being the greater drift hazard associated with irregular shaped, buoyant dust particles. If a 20μ particle is allowed to fall through a height of three metres, and subjected to a gentle breeze of five kilometres per hour, it can be calculated that the horizontal movement of that particle will be more than 350 metres. Drift results in application failures and in damage to adjoining crops. The application of dust formulations, particularly from aircraft, is thus limited to a much more restricted range of climatic conditions compared with the use of sprays, and the necessary safety precautions may be much more stringent.

The drift problem ceases to exist if the insecticide is formulated as the larger granules, rather than as a dust. Retention of an even deposit over a leaf surface is of course impossible with granule formulations and they are used only when this is unnecessary or undesirable. For example, granules containing a systemic may become lodged in the whorls of a plant's foliage and gradually give up the insecticide to the plant, giving a more extended period of protection than is the case with a systemic spray. They are very useful for soil treatment at the base of plants with dense foliage, such as forest and fruit trees, particularly if aircraft are to be used for their application, since their high powers of penetration ensure that very little insecticide is filtered out by the leaves. One of the early uses of granules was in the treatment of marshes for mosquito control, where dense reed growth prevented much of a normal spray or dust from reaching the water. It is now common practice to modify seed drilling machinery so that insecticide granules can be placed in the soil at the time of crop sowing. The slow release of poison ensures the protection of the young growing plant at the period when it is most susceptible to attack by soil pests. Furthermore the precision with which the granules are placed in the rows eliminates the waste of material which occurs in 'blanket' treatment with other types of formulation, and this offsets the rather greater cost of granular materials.

At the other end of the particle size range are the smokes, which consist of very small and buoyant solid particles which will remain airborne for considerable periods of time. These are widely used in situations where pest control necessitates filling an enclosed volume of air with insecticidal particles, such as on crops in glasshouses, against stored products pests in ships and warehouses, and against pests of structural timbers. The smoke particles gradually settle out to form an active layer over all horizontal, or near horizontal surfaces. Smokes are produced by generators in which the insecticide formulation contains heat producing combustible materials, which are sealed usually in a cardboard or aluminium drum. The mixture is ignited and burns, without flame, to release a dense cloud of particulate pesticide.

The most commonly used insecticidal formulations are designed for spraying by machines of the types described in the next chapter. The liquid used as a carrier for the insecticide must be non-phytotoxic, inexpensive and readily available wherever the spraying is carried out, and therefore the obvious choice is water. Very many organic pesticides are solid substances, comparatively few of which are soluble in water, and the first formulation problem is to obtain a homogeneous spray fluid. One method is to disperse fine particles of the active ingredient in water and spray the resulting suspension. One of the main problems with suspensions is to prevent the particles settling out as a slurry or sludge in the bottom of the spray tank, a process which obviously results in exceedingly uneven dosage on the crop. This tendency can be partially prevented if an efficient mechanical agitator is incorporated into the design of the tank, but more important is the introduction of auxiliary substances known as dispersing agents into the pesticide formulation. These agents tend to prevent sedimentation by forming a protective shell around each particle on to which they are adsorped which prevents the particles coalescing. They also raise the viscosity of the solution, thus decreasing the rate of settling.

Apart from settlement problems, the main difficulties with suspensions arise when high concentrations are required, or when fine nozzle apertures are used. The probability of the particulate matter blocking nozzles increases, causing inefficient application and wastage of time and money, which, in the case of aircraft applications, for example, could be serious. Generally, therefore, completely liquid spray formulations are preferred and since most pesticides are insoluble in water, the most widely used method of application is to spray an emulsion. An emulsion is formed when two immiscible liquids are shaken together so that small droplets of one of the liquids are suspended in the other. In order to produce a pesticide emulsion the active ingredient must first be dissolved in a solvent, which is usually mineral oil. Solvents are an important category of auxiliary substances in that factors such as their viscosity, volatility and surface tension have an important effect on the type of spray produced and the manner in which it is deposited. A viscous oil with a high boiling point will tend to produce large droplets which settle rapidly, whereas a very volatile oil with a low surface tension will tend to produce small droplets more suitable for spatial sprays directed against flying insects. The mineral oils which are normally used as insecticide solvents do not form stable emulsions when mixed with water. On standing, the droplets coalesce and the mixture becomes separated into oil and water layers. Consequently, another important class of auxiliary substances, the emulsifying agents, are incorporated into the formulation. Many substances can act as emulsifying agents since they consist of molecules which can become orientated around oil droplets to form a surface barrier which will prevent coalescence. Examples are the calcium salt of milk protein (calcium caseinate) and soaps (sodium salts of long chain fatty acids).

As in the case of dusting, one of the perennial problems with spray applications is the drift hazard. This is particularly acute in the case of aerial spraying and imposes strict limitations on the range of wind and evaporation conditions suitable for spray operations. Any solution to the drift problem would mean more flying hours per day for the aircraft and therefore lower application costs. One technique which has been patented is the use of very viscous carriers in place of water in the spray formulation; the high viscosity reduces the proportion of small droplets in the spray, and various auxiliary substances are included to reduce evaporation. It is also possible to distribute aqueous droplets, containing the active ingredient, in a viscous oil carrier; in other words to produce an invert emulsion of water in oil, instead of oil in water, the drift hazard with invert emulsions being considerably lower than with normal emulsions.

The main modern development in spray technology is the production of machinery which will efficiently deliver much higher concentrations of insecticides in a much lower volume of liquid (see section 5.2). The large bulk of water can be reduced and in some application procedures the operator can dispense with water altogether. In these cases the sprays are formulated as true solutions, a good example being the large amounts of DDT which have been applied as oil solutions from aircraft. The main formulation consideration that applies to true solutions is the choice of a suitable solvent which will provide a solution having as low a cost and phytotoxicity as possible.

No matter what spray procedure is adopted it is essential to ensure a lethal deposit on the target, and this involves further problems which auxiliary substances can help to solve. If a water droplet is placed on the waxy cuticle of an insect or the leaf of a plant it will frequently roll off, leaving the surface dry, and this must be prevented if a continuous film of insecticide is to be deposited on the surface. Auxiliary substances which enable the insecticide to spread over the treated surface and form a stable layer of liquid on it are known as 'spreaders' and 'wetters' and are invariably included in spray formulations. Originally soaps were used as spreaders, but the precipitation that occurs with hard water imposes a limitation on their use. At the present time, many of the synthetic detergents are used as spreaders and wetters.

In the case of insecticides applied to plant surfaces, one of the important factors is the adhesion of the deposit on the surface, and the tenacity with which this deposit will resist weathering by dew and rain. Auxiliary substances known as 'stickers' are occasionally included in both dust and spray formulations to improve retention and persistence.

Spraying is not the only method of applying insecticides in liquid form (see Table 11). If the liquid droplets are below 30μ in diameter they are said to form a fog rather than a spray, and fogs are useful for similar space-treatment problems to those described in connection with smokes.

The production of these small droplets depends upon choosing solvents with suitable boiling points and viscosities, and applying them with specially designed aerosol generating equipment. There is close overlap here between the effects of formulation and those due to the equipment used and the topic of aerosol production is discussed in detail in the next section (5.2.2).

The other liquid application techniques described in Table 11 require no atomisation equipment. Each technique raises its own formulation problems, whether they be the penetration of a fleece by a sheep dip, the impregnation of timber to control wood borers, or of textiles to control clothes moth, or the introduction of a systemic into the irrigation water of a glasshouse. These together with the formulation techniques for the preparation of insecticidal paints, resins, papers and other surfaces (Table 11), are in many cases highly specialised problems and their detailed discussion is beyond the scope of this generalised account of formulation.

In the case of vapour applications, or fumigation, the penetration and effectiveness of the insecticide depends far more upon physical factors such as temperature, humidity and time of exposure, than upon use of auxiliary substances. Very frequently the pure gas is used, either being released from cylinders where it is stored under compression, or by generation on the spot from volatile liquids, or solids, or by a chemical reaction. The only formulation requirements are specialised ones such as the mixing of a relatively harmless gas with a fumigant of low flash point, in order to reduce fire and explosion risks.

5.2 Application

There are other factors, in addition to the correct choice of formulation, which affect the efficiency and cost of chemical control techniques. Often of considerable importance is the application technique adopted, the effectiveness of which depends upon the skill and experience of the personnel concerned, and the suitability of the machinery available to them. Skill and experience should ensure that the correct poison, formulation and dose rate are applied at the correct time to produce the maximum effect on a vulnerable stage in the pest's life cycle, with the minimum of toxic side effects. But the skilled worker may require specialised equipment if the full potential of the control measures is to be realised. The complexity of this equipment varies, in a similar way to chemical formulation, according to the type of pest problem involved. Thus, if a pesticide is being administered to farm animals in their drinking water, no specialised equipment is required, and if bait is distributed for a localised grasshopper control programme, the equipment involved may be no more sophisticated than a shovel. On the other hand, if the pest is very active and cannot be lured to a bait, or if one is attempting to cover with poison all possible surfaces on which the pest may move or feed, complex equipment may be required to ensure an adequate distribution of the chemical.

The choice of this equipment is determined by the formulation to be used, to the extent that machines for applying solid materials generally differ in design from those used for applying liquids.

5.2.1 *Equipment for applying solid formulations*

(a) Dusters. The long history of use of inorganic pesticides has been described in section 3.2.2(a), and until the end of the last century these were always applied as dust formulations. Dusting is therefore the original method of applying insecticides to crops, and it is generally easier to carry out than spraying, in that the active materials are lighter and less bulky, and a given area can be treated with dust more rapidly than with a spray. Nevertheless, since the 1880s, the development of dusting equipment has lagged far behind that of spraying equipment, the reason being that certain fundamental difficulties associated with dust applications have never been satisfactorily resolved. These difficulties include the greater drift and retention problems mentioned in section 5.1, and also the fact that materials tend to separate on application, uniform deposits of mixed dusts being difficult to achieve. Thus one spray may contain two or more insecticides and fungicides mixed together and as long as the chemicals are compatible (do not react with one another), a single spray application may serve several crop protection purposes. In the case of dusts each chemical usually requires a separate application. Another difficulty is that of ensuring an even rate of delivery from a dusting machine. Dusters consist of a container, which releases dust into an air stream, which in turn carries it to some sort of nozzle. Many of the older dusters suffer from the disadvantage that as the amount, or 'head' of dust, in the container falls, the rate of delivery shows a corresponding decline. One comparatively modern development is the incorporation of some sort of conveyer belt or metering device to ensure a constant delivery rate in large capacity dusters.

Dusters can be classified on the basis of their size and mode of operation into four groups: hand dusters, which normally contain only a pound or two of dust which is expelled by the hand operation of a simple bellows or piston; knapsack dusters, the containers of which usually contain 2–9 kg of dust and are carried, slung from the operator's shoulders; traction dusters, in which the container is drawn along on wheels, these being geared to a fan which provides the flow of carrier air; and power dusters, in which the fan is driven by an independent engine. These types of machine are illustrated in Fig. 13 and further details are provided in the caption to this figure.

Most modern large-scale dusters fall into the fourth category, in which the fan is usually driven by a petrol engine, although occasionally the power take-off of a tractor is used. These power dusters are basically of two types, those in which the dust laden air flow is divided up to supply several nozzles, and those in which it is ejected from one or two large orifices. The former type is used for dusting arable crops grown in rows (row crops). The nozzles are fixed at regular intervals along a horizontal boom, attached to the rear of a tractor. The spacing of the nozzles corresponds to the spacing of the crop rows, and each nozzle is connected to the dust distributor head on the machine by a flexible pipe (see Fig. 13). Frequently an enclosing hood is fitted over the boom in an attempt to reduce drift. By contrast, if a single orifice duster is employed, the dust may be deliberately ejected upwards into the air and left to settle over the crop on the downwind side of the machine. This technique of 'drift dusting' involves the machine in fewer, more widely spaced journeys across the area to be treated, but the attainment of a sufficiently even distribution is an obvious problem. The main use for single or double orifice 'dust blowers' is the treatment of trees, and very powerful fans have been developed which produce high velocity air flows capable of projecting a column of dust to the tops of even large forest trees. However, the retention problems already discussed in connection with dust deposits (section 5.1) became accentuated if powerful blasts of air are being used, and various attempts have been made to overcome these problems. One partial solution is to arrange spray nozzles around the dust outlet and coat the particles with oil or water as they leave the duster. This 'spray-dusting' technique is quite effective, but when the dusting process reaches this level of complication one is forced to ask whether a straightforward spray application would not be a feasible alternative. In fact one of the main contributions of air blast dusters to the technology of pesticide application is that they have served as stepping stones in the development of spray blowers (see below). Another suggested means of solving the deposition and retention problems is the use of an 'electrodusting' technique.

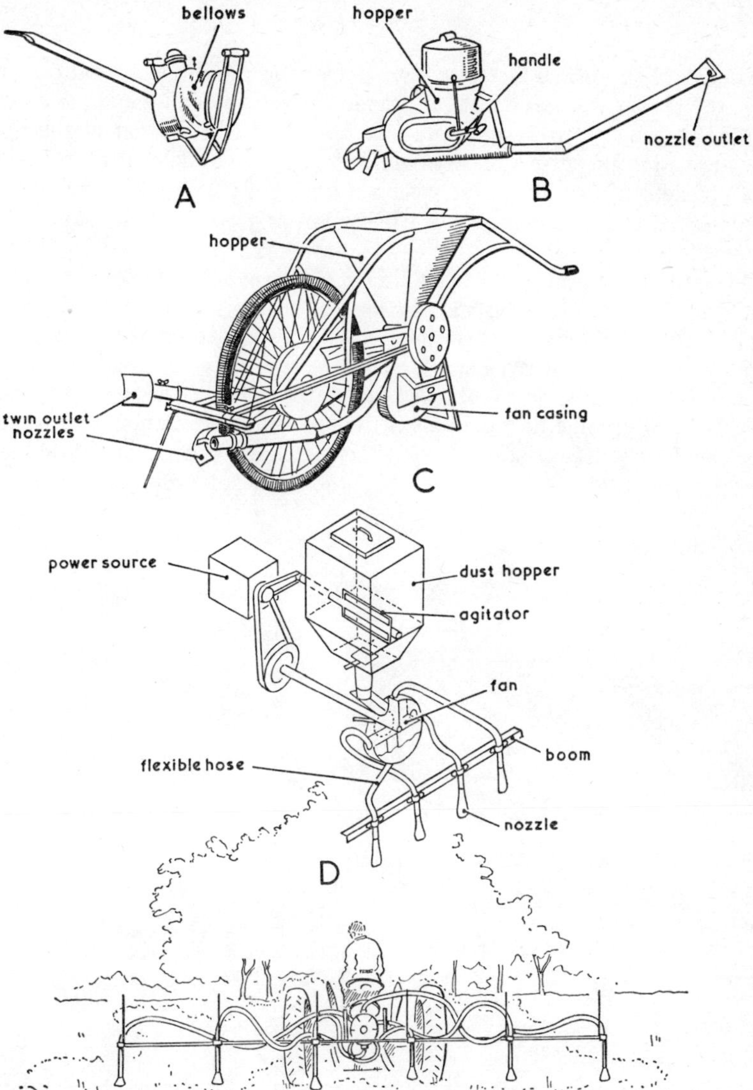

Fig. 13 Dusting equipment.
A. Hand duster. The air blast is provided by a hand operated bellows.
B. Rotary duster. These dusters are of the knapsack type, the hopper being worn on the back or chest of the operator. The air blast comes from a hand operated rotary fan.
C. Wheelbarrow duster. This is a traction duster, the motive power for the fan which provides the airblast coming from the single large wheel in contact with the ground.
D and E. Power duster. All large scale treatments utilise airblast fans driven by an independent motor. This shows a row crop duster in which the air blast from a centrifugal fan carries dust through flexible pipes to nozzles arranged along a boom.

This depends upon the fact that plant surfaces carry an electrostatic charge. If, as the dust leaves the nozzle, it passes between charged metal plates, each particle may be given an opposite electrostatic charge to that of the plant, thus increasing the attraction between dust particle and leaf surface. Laboratory and field trials, carried out over a number of years, have demonstrated the potential of this technique, but it has not as yet found wide commercial exploitation.

(b) Granule applicators. Other solid formulations require specialised application equipment. Granules may be applied to row crops by hand, using a small hopper and delivery pipe fitted with a simple metering device to ensure a constant rate of delivery. On a larger scale they may be broadcast, using a rotary spreader of the type employed for fertilisers. However, one of the main advantages of granule formulations is the precision and economy with which they can be used (section 5.1) and Fig. 14 shows a seed drill fitted with

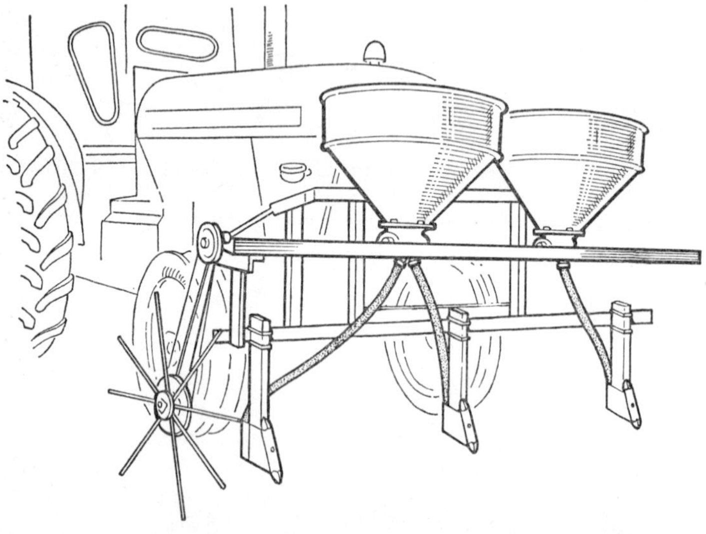

Fig. 14 A tractor mounted applicator which can be used for placing a band of insecticide granules in the vicinity of plants at the time of sowing. Flexible pipes carry the granules from the two hoppers to the coulters which place the granules in a band which is aligned with the share of the planter.

granule hoppers from which these particles are metered into the soil above the seeds.

(c) Smoke generators. Smokes are usually applied from burning cartridges of the type already described in section 5.1.

5.2.2 Equipment for applying liquid formulations

(a) Spraying machines. The technique of applying crop protection pesticides as liquids dates from Millardet's discovery of the fungicide Bordeaux mixture in 1882, and since that time it has become the main method of releasing pesticides into the air or depositing them on plant and other surfaces. The aim of any application technique is to ensure lethal contact with the pest, and if the location of the pest cannot be accurately pinpointed, this aim is achieved by distributing an adequate quantity of pesticide as evenly as possible over the infested area. The application of a liquid pesticide to a surface could be achieved by using a jet from a hose, but the jet delivers too large a volume to too small a surface area, and is therefore very wasteful and probably damaging. If the liquid jet is broken into droplets which are dispersed as a cone or fan-like spray as they leave the nozzle, a much greater area of the surface is treated, and a much higher proportion of the liquid is retained on the surface in unit time.

Spraying machines can be classified on exactly the same basis as dusting machines; hand sprayers, for very small scale applications; knapsack sprayers, which one man can use to apply up to 25 litres of spray fluid for treatment of larger areas; traction sprayers, which were usually horse drawn and utilised the now obsolete method of driving the pump by gearing from the land wheels of the sprayer; power sprayers, a group which includes all of the modern large-scale equipment in which the pumps are driven by an independent power supply. Sprayers can also be classified into pneumatic and hydraulic machines depending upon whether the liquid is forced to the nozzle outlet by pressurised gases or by a hydraulic pump. Further explanation and illustrations of these various types of equipments are given in Fig. 15.

The spraying machine is designed to produce a cloud of droplets which will, ideally, impinge on the surface to be treated, spread out and leave it uniformly wetted. In the past, the method of ensuring this uniform distribution of active chemical was to dilute it with large amounts of water and spray until the treated surface was saturated and liquid began to drip from it; this is termed spraying to 'run-off'. Various types of nozzles will provide the atomisation necessary for this type of application, a widely used example being shown diagrammatically in Fig. 16.

The liquid on reaching this nozzle, is forced through the swirl plate. This is simply a flat plate, pierced obliquely by two or more holes, so that the liquid is diverted and leaves the swirl plate as two or more threads, at an angle to its former forward path. These threads strike the wall of the eddy chamber obliquely and the liquid is therefore given a rotary motion. Thus, on leaving the nozzle aperture, it is subject to two forces, one acting forwards, due to the pressure of the liquid in the pipe, and one actually laterally, due to the centripetal force imparted by the swirling in the eddy chamber. It therefore spreads out as an expanding conical sheet of liquid, which is torn into droplets by the frictional forces due to the surrounding mass of inert air.

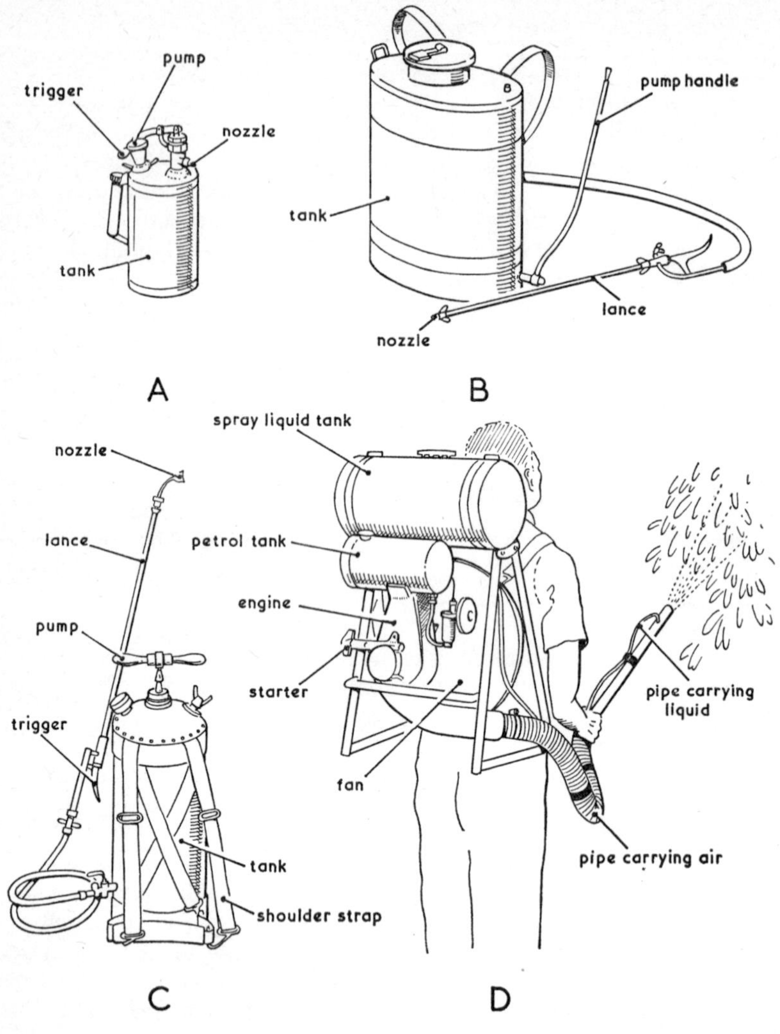

Fig. 15 A. Hand sprayer. This operates pneumatically, the pump compresses air above the liquid in the tank. On releasing the trigger the air pressure forces liquid through a feed pipe to the nozzle.

B. Knapsack sprayer. Slightly larger scale spraying operations require lrger quantities of liquid which are carried in a tank worn on the operator's back. The sprayer illustrated here is a hydraulic sprayer. Continuous pumping by the operator forces the liquid along the pipe to the nozzle. The tank is not pressurised.

C. Knapsack sprayer. The tank here is pressurised with air and this belongs to the group of pneumatic sprayers together with A, above. After pressurising with the pump the sprayer will operate for ten minutes or more without requiring further pumping. Pneumatic sprayers therefore impose less strain on the operator although the tanks and hoses are subjected to greater strain.

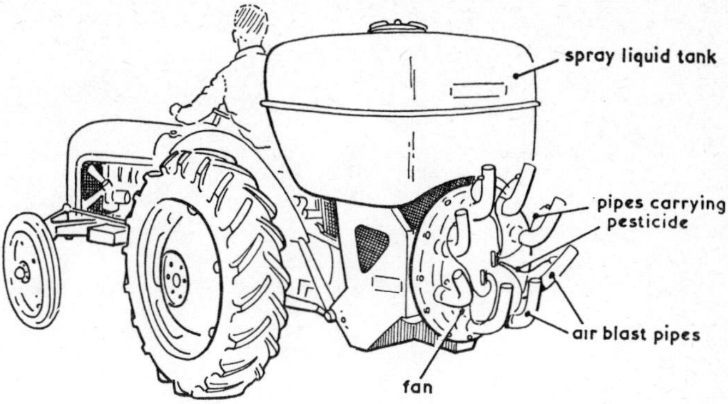

E

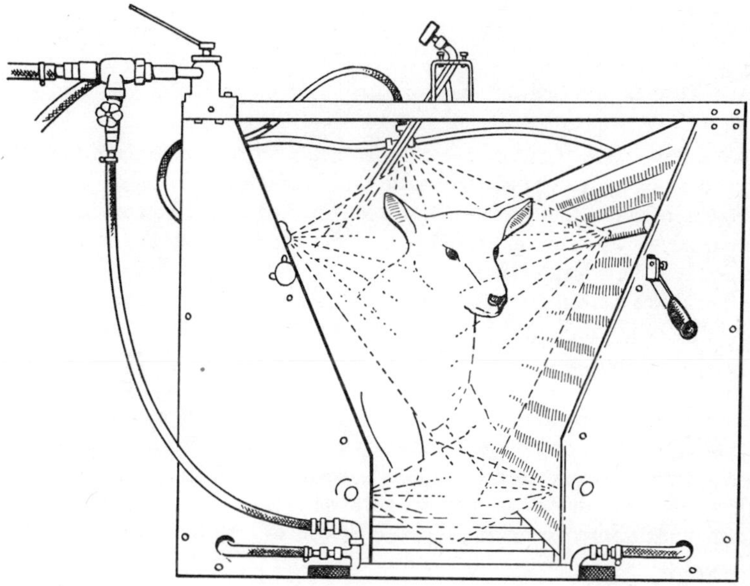

F

D. Motorised knapsack sprayer. For large scale work an independent engine or the power take-off of a tractor is used to hydraulically pump liquid to the nozzle. This shows a knapsack sprayer fitted with its own motor. Atomisation is assisted by an air blast from a fan which is also driven by the engine.

E. Tractor-mounted air blast sprayer for orchard work. The power take-off of the tractor is used to drive the fan providing the air blast and also to pump liquid pesticide to outlets placed in the path of the air blast.

F. Spray race. Batteries of hydraulic nozzles are arranged so that sheep passing through the spray are thoroughly saturated with pesticides suitable for ectoparasite control.

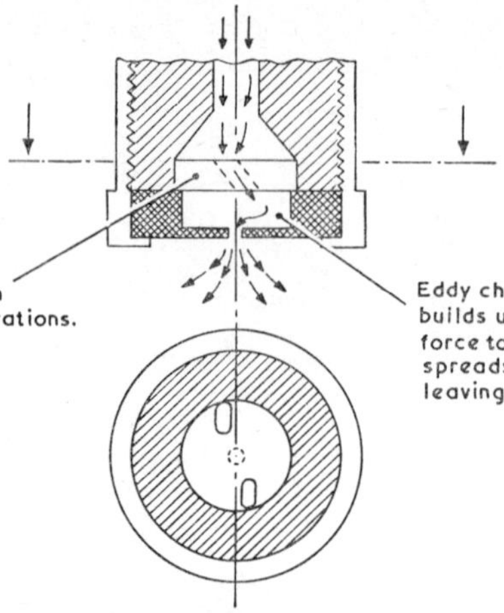

Fig. 16 Diagram showing the working parts of a generalised hydraulic nozzle.

This 'hollow cone' pattern of spray droplets can be converted to a 'solid cone' type by using a swirl plate pierced with a small central aperture, which allows some of the liquid to follow a direct forward path. 'Fan nozzles' have the actual nozzle aperture elongated into a slit lying at the base of a groove. The spread of the spray is thus restricted to one plane and the spray pattern consists of a flat fan of droplets. Each of these nozzles can be fitted along a horizontal boom, the height of which is adjusted so that the cones or fans meet to give maximum coverage of the rows of the crop being treated.

However, the importance of nozzle design is not just a matter of varying the shape of the spray cloud. The type of nozzle used partly determines two more fundamental characteristics of the spray, the mean droplet size, and the range of droplet sizes produced, usually called the 'droplet spectrum'. When an object, for example a twig, is placed in the path of a moving spray cloud, the frontage of approaching air, equal in width to the diameter of the twig, divides and passes as 'streamlines' on either side. Only droplets in a narrow central strip of the approaching front are certain to impact with the twig, those on either side are liable to be carried round in the streamlines of the moving air boundary. Penetration of this boundary layer is possible in the case of large particles with sufficient inertia to resist the change of direction of the stream lines. Smaller droplets will only impact on the twig if they have a high enough velocity to traverse the moving boundary layer before it can carry them away. The percentage of the droplets in the approaching front which is removed by the twig is known as the 'dynamic catch', and is very much affected by the type of spray machine used. If, for example, a heavily forested area is being sprayed to control flying mosquitoes, then a low dynamic catch is required; the spray must remain airborne and not be filtered out by the foliage. Relatively slow moving droplets of small diameter (30μ or less) will meet these requirements and the appropriate formulation and machinery is selected. For crop spraying, on the other hand, a high dynamic catch is essential, since a large percentage of non-impacting droplets increases the potential drift hazard. The ideal droplet size is therefore a compromise between the small size required for a high degree of dispersion, and the large size required for a high dynamic catch, and a hydraulic nozzle which produced only droplets in the 90–160μ range would meet these requirements perfectly. However, the droplet spectrum of nozzles of the type described above is usually very much wider than this and the production of drift droplets of 30μ diameter, or smaller, is inevitable. One can alter the nozzle design so that the whole spectrum is shifted and no drift droplets are produced, but the average drop size is then so coarse that the spray is useless for pesticide application. When hydraulic nozzles are used drift must therefore be tolerated, but can be reduced by the use of hooded machines (as in the case of dusters) and sometimes by modifications to the pesticide formulation (see section 5.1).

As already pointed out, adequate distribution can be ensured by high dilution of the material applied, and the consequent application of large volumes of spray fluid. Applications of 1100 l per ha on ground crops are commonplace, and on fruit trees the volumes required are much higher, up to 11 000 l per ha in the case of treatments on dense citrus foliage. The high pressures (up to 7MN/m^2) required to reach the tops of trees with large volumes of liquid impose considerable strain and wear on machinery and considerations of this sort, coupled with drift problems, inspired the search for other means of atomising spray fluids.

It can be shown that the production of droplets of a given size is influenced by several factors. Droplet size is directly proportional to the viscosity of the liquid and inversely proportional to its density; it is also affected by the surface tension, and all of these factors can be altered by changes in formulation. In addition, droplet size is inversely proportional to the relative velocities of gas and liquid, and the relative volume flows of gas and liquid, at the nozzle outlet. Now, in the case of the nozzles described above, the volume and velocity of the gas (air) flow is nil, atomisation occurring as a result of friction between the moving liquid and the stationary air. Therefore, if the air is moved over the nozzle outlet, increasing the relative gas : liquid flow, smaller droplets will result. Conversely, the same droplet size can be retained, but can be produced by a much coarser hydraulic nozzle, the effect of moving air being to increase the degree of atomisation. If coarser nozzles can be used, higher outputs are possible, without imposing upon the equipment the strain of higher pressures.

These arguments have led to the development of a variety of 'air-blast sprayers' in various parts of the world, in which a coarse low pressure spray is smashed into smaller droplets when ejected into a powerful blast of air. The airflow also directs the droplets towards the target, thus reducing the drift hazard. This basic concept is not a new one; inventors have attempted to exploit air flow techniques almost since spraying started in the 1880s, but various technical difficulties, such as the lack of suitable formulations, hampered development in this field.

Several attempts were made in the 1920s and early 30s, particularly in America, to produce a completely successful spray blower. The first high output machine to be extensively marketed was the speed sprayer, developed at the end of the 30s. It consists of a trailer tank, at the rear of which is an axial flow fan which ejects air at the rate of several thousand cubic feet per minute through a wide arc shaped outlet, around which from 42 to 150 hydraulic nozzles are arranged. Hydraulic pressures as low as 350–500 kN/m^2 provide adequate atomisation.

At the present time the successful distribution of large amounts of spray fluid is possible with any one of a large number of spray blowers available to orchard growers (see Fig. 15). This, however, is not the only benefit conferred by the development of air atomisation techniques. In fact increases in the quantity and complexity of applications result in increases in the expenditure of time and money, and the development of air blast techniques raised the possibility of replacing much of the water used as a spray carrier by air, and thus reducing the volume of liquid required. The realisation of this possibility has altered the whole outlook on spray application in the last twenty years.

The basic reasoning in the development of what are called 'low volume' spraying techniques is as follows. Run-off spraying is necessary to achieve effective distribution with the old-fashioned wide spectrum hydraulic nozzles, but if a very even distribution of uniform size droplets can be achieved the surface can be effectively covered by far less liquid, and thus with far less expenditure of time and effort. The droplet characteristics become critically important. If a spray is applied evenly at 11 l per ha, and all the droplets are of 1000μ diameter, there will theoretically be 1 droplet on every 6 sq cm of the flat sprayed surface. If, on the other hand the same rate of application is used with 5μ droplets there will theoretically be about 9 million droplets per 6 sq cm. Obviously each 6 sq cm of surface is going to be more adequately covered by nine million small spots, than by one relatively large splash. If the volume of spray solution is reduced, the concentration of pesticide increases and the phytotoxic and other hazards presented by the excessively large or excessively small (drift) droplets of the orthodox wide-spectrum nozzles cannot be tolerated. Air atomisation can produce a much narrower droplet spectrum, and furthermore the increased velocity imparted to the droplets by the airflow increases the dynamic catch of smaller droplets (see above).

Experiments were started in Canada in 1946 in which an airblast sprayer was used to spray 600–1200 l per ha of concentrated pesticide onto fruit trees, in place of the 4800 l per ha of dilute spray which the machine normally delivered. Successful control was achieved, together with a substantial saving on labour and chemical costs. Experiments started in Britain in 1947, using the fungicide lime-sulphur against apple scab, showed that very high concentrations of pesticide could be used without damage to the tree, provided that effective atomisation was possible. Concentrations of 100% pesticide were applied at the ultra-low volume of 30–70 l per ha by means of a paint sprayer, in which compressed air is used to produce a very narrow droplet spectrum.

Thus the air blast machines, originally designed for very high outputs, were as it were pre-adapted for the new low volume techniques since the droplet size and spectrum could be controlled by regulating the air and liquid flow rates. Subsequent development has involved a search for improved nozzles and some machines, which can be classified as 'mist blowers', do not atomise the liquid with the orthodox type of hydraulic nozzle. Mist blowers rely entirely upon the air blast to produce the necessary atomisation (except in the case of the rotary atomisers—see below) the simplest method being to feed the liquid from an open pipe into the air blast. This type of 'shear nozzle' was at one time used for feeding liquid into the slipstream of spraying aircraft, and small versions are employed on some tractor drawn and motorised knapsack mist blowers. The open end of the shear nozzle is usually cut obliquely to form a sharpened orifice of oval cross section. Droplets are sheared from the rim of this orifice by air flows of up to 400 km/h, and these air flows, or a lower velocity auxiliary flow, carry the spray to the target. Other mist blowers have been designed in which the liquid strikes a fixed metal plate or cap, from the rim of which it spreads out as a sheet into the path of the air blast which atomises it. The output of these mistblowers can be varied over wide limits, the droplet size, which is usually smaller than that produced by spray blowers, being controlled by the relative flow rates of air and liquid. There is a third class of mistblower however, which is purely for low volume application. These machines are the rotary atomisers, which consist of rotating discs or wire gauze cylindrical cages into the centre of which the spray liquid is fed. Droplets of very even size are formed as the liquid is flung from the rim of the disc or the meshes of the cage. The size is controlled by the speed of rotation and the flow rate of the liquid. Rotary atomisers are widely used on crop spraying aircraft, being driven by a windmill placed in the slipstream (see Fig. 17). For ground application a comparatively low power motor is required to rotate the atomiser, and usually a fan provides an airstream to carry the droplets to the crop, although some recent machines rely entirely upon the centripetal force to fling the droplets onto the plants. There is an upper limit to the amount of liquid that the rotary atomiser can cope with, but for low and ultra-low volume applications these machines are highly successful by virtue of their very narrow droplet spectrum.

The low volume techniques, utilising spray blowing and misting machines, which have been described so far, are predominantly suitable for fruit tree

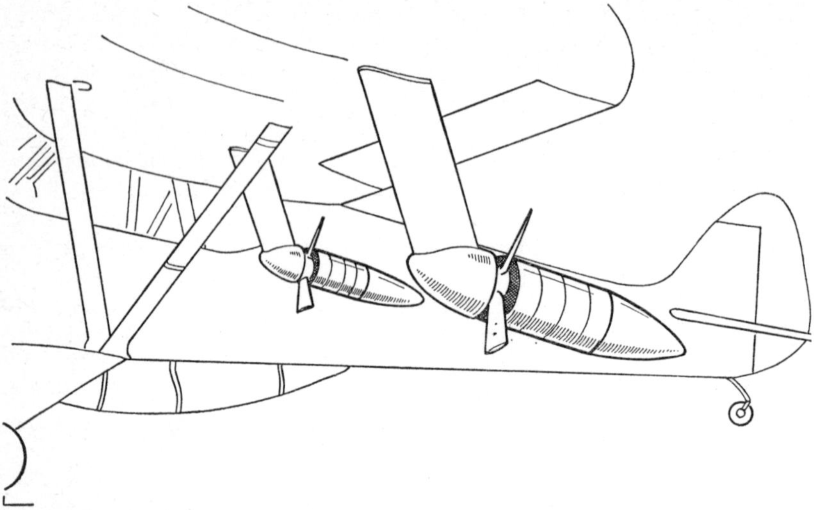

Fig. 17 Rotary atomisers mounted on an aircraft. The slipstream turns the propellors which rotate a battery of spinning discs on to which insecticide is fed. Uniform droplets are ejected centrifugally from the edges of the discs.

treatments. Various attempts have been made to modify the air blast principle in nozzles for attachment to booms for the treatment of row crops. However, most low volume boom sprayers rely upon other nozzle designs. In the late 1940s American workers discovered that simple fan jets of the type fitted to air humidifiers gave good atomisation at low hydraulic pressures when fitted to a crop spraying boom, and moulded ceramic gas jets have subsequently been widely used.

The speed, efficiency and economy of low volume spraying have resulted in its widespread use at the present time. However it must be pointed out that not every insect pest is controlled as effectively by low volume as by high volume techniques. Furthermore, the small droplet sizes and high concentrations increase the potential drift hazard, and some of the most dangerous pesticides, with the highest mammalian toxicities, cannot be sprayed at low volume at all.

(b) Aerosol generators. It was seen above that for a pesticide particle or droplet to remain airborne for a long period of time, the average diameter should not exceed 30–40μ. A pesticide at this level of dispersion is referred to as an aerosol, and if solid it is known as a smoke, if liquid it is called a fog (see section 5.1). Liquid aerosol droplets can be produced by equipment working on the principles already discussed, e.g., very high speed air jets, or very rapidly rotating discs. However, many of the aerosol generators on the market differ in important respects from other spraying equipment. It has been shown that droplet size is proportional to the viscosity of the spray fluid. This viscosity can be affected by altering the formulation, but if very small droplets are required a much greater lowering of the viscosity can be obtained, by the application of heat. The machines known as 'thermal aerosol generators' therefore atomise the liquid with a blast of heated air, and the droplet size is further reduced by the initial evaporation of the oil carrier as the droplet is produced.

The machine known as TIFA (Todd Insecticidal Fog Applicator), was developed from wartime naval smoke screen equipment. Air is passed through a combustion chamber and heated to 480–600°C (900–1100°F); insecticide is fed into the base of a cup around which it is swirled by part of the heated airflow, and it is then flung from the rim of the cup into the main flow and atomised. Condensation at the aperture produces a dense fog.

More portable machines have been devised to utilise the hot exhaust gases of aircraft or motor vehicles, but one of the most widely used of the portable generators is 'Swingfog', a German invention which utilises a pulse jet engine, a large scale version of which powered the V-1 flying bomb in World War II. The insecticide is dissolved in a light oil and fed through a shear nozzle into the exhaust pipe of the pulse jet (see Fig. 18). The explosion of the engine charge drives the mixture forward, further mixing occurring with the cool air flowing in the outer pipe.

The insecticide vapourises and condenses beyond the outlet aperture forming a dense fog of fine particles, the particle size of these being controlled by adjusting the insecticide flow.

Other thermal aerosol generators depend upon the energy of super-heated steam for fine atomisation. Steam and insecticide are pumped separately to a nozzle where they are mixed, and the very high expansion velocity of the steam converts the insecticide into a fog.

Thermal aerosol generators are used very widely in the world today against pests of medical importance, such as mosquitoes, blackfly and tsetse, against orchard and forest tree pests, and in enclosed spaces against glasshouse and stored products pests. Their outdoor use on agricultural crops is obviously limited by the small droplet size but they are employed in locust and grasshopper control programmes and have even been used for fogging ectoparasites on sheep.

The aerosol generators which are most familiar at the domestic level, for the control of household pests such as flies and cockroaches, do not of course work on the thermal principle. In these the insecticide is formulated as a solution in a low boiling point liquefied gas. This is held under pressure in a container fitted internally with a capillary tube leading to a fine atomising nozzle at the top. The vapour pressure of the liquid provides the force to drive the insecticide solution up the capillary when a trigger is released. The liquid boils and evaporates, leaving small droplets suspended in the air beyond the nozzle. The propellants employed are usually halogen derivatives of simple paraffins such as dichlorodifluoromethane (Boiling Point −29·8°C),

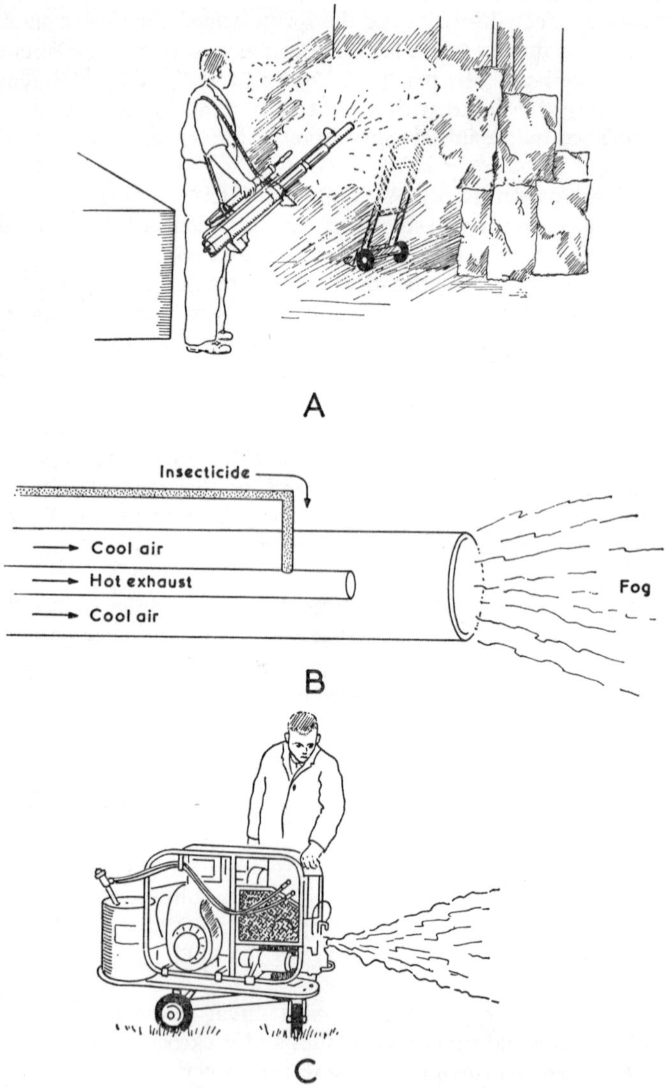

Fig. 18 Thermal fog generators

A. Swingfog. The insecticide solution is fed into the hot exhaust of a pulse jet engine.

B. Diagrammatic representation of the nozzle arrangement in a swingfog type of generator.

C. T.I.F.A. The Todd Insecticidal Fog Applicator, in which the heated exhaust is obtained by burning petrol in a combustion chamber.

trichloromonofluoromethane (B.P. 23·7°) and 1:1 difluoroethane (B.P. −24·7°). The choice of propellant depends upon the type of aerosol generator, or bomb, which is used. Strong metal bombs, for use in glasshouse work, can withstand the high vapour pressures characteristic of the low boiling point propellant, whereas the more fragile tin or plastic containers commonly sold for domestic use must operate at lower pressures. Accordingly higher boiling point materials, such as trichloromonofluoromethane (see above) may be included with the low boiling point compounds to produce a lower vapour pressure mixture. The vitally important factor, where household use is concerned, is to employ propellants and insecticides which are virtually non-toxic to mammals, pyrethrum mixtures being very popular in this field of insect control.

(c) Other application equipment. Spraying and dusting techniques are used on a wide scale in agriculture, forestry, horticulture and medicine for the application of a great range of pesticides, and these techniques have therefore been discussed at length. Only brief mention can be made of some of the application techniques which, although important in their particular field, are nevertheless restricted to more specific pest problems. Spraying and dusting are necessary when the precise position of the pest in space or time is in some doubt, but this is not always the case. If the pests are, for example, internal parasites of animals, the pesticide can be applied with much greater precision.

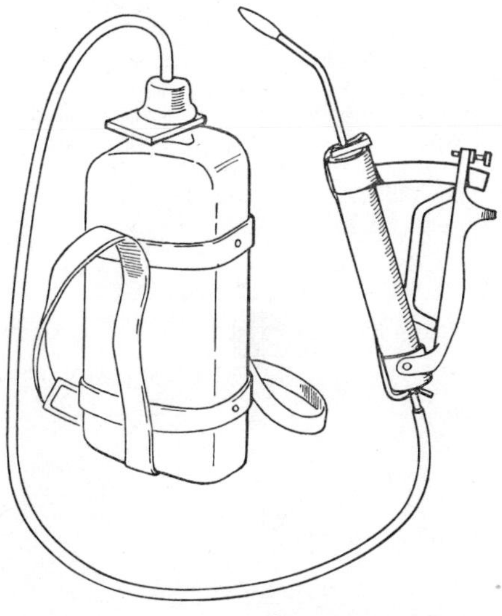

Fig. 19 A knapsack drenching gun for applying anthelmintics to sheep. The tank containing the pesticidal solution is worn on the operator's back; the nozzle is pushed into the sheep's mouth and operation of the trigger forces a measured, pre-set dose down the animals throat.

Thus gut parasites can be controlled by a specified dose of chemical which, when administered orally to the host animal, will almost inevitably make contact with the pest. A variety of dosing guns, each of which can be connected to a knapsack supply of pesticidal liquid, are available for the rapid precision dosing of large numbers of host animals (see Fig. 19). A somewhat analogous situation may occur in the application of systemics to plants. A meter can be used to mix a precisely known concentration of the insecticide into the irrigation water for glasshouse plants. Good distribution is possible in this case with a simple piece of equipment, and the same is generally true of vapour applications, where the physical characteristics of the pesticide ensure a uniform dispersion using equipment which usually simply consists of cylinders of liquefied gas.

Even more specialised application techniques are required for problems such as the impregnation of wood and textiles with insecticides, but space does not permit a discussion of these restricted application problems.

Chapter Six
CONTROL OF ANIMALS WITH PEST STATUS BY POPULATION REDUCTION: Biological Methods

6.1 Direct Reduction Using Organisms of Other Species

6.1.1 Introduction

The population of any species of animal does not increase indefinitely in size but tends to reach an optimum level about which it fluctuates. This concept of the limited size of animal populations, originated by Malthus in 1803, played a significant role in Darwin's formulation of the theory of organic evolution. The tendency of the environment to limit increases in numbers is termed natural control. Biological control, which is one factor in natural control, has been defined by De Bach (1964) as "The action of parasites, predators or pathogens in maintaining another organism's population density at a lower average than would occur in their absence".

Population densities of pests are largely determined by climate and the activities of other living organisms. Generally speaking the more heterogeneous a community of animals is the more stable are the densities of its member species. Conversely homogeneous communities tend to lack stability and this is often the reason why pest outbreaks occur in environments where natural predators and parasites are rare, a situation which may arise when pests are introduced into new environments (see section 1.3.3).

Applied biological control is the use of living organisms by man to reduce or at least limit pest population size. Because of the density dependent nature (see section 1.4.1(b)) of biological control agents they usually act continuously and since they are self perpetuating, once established are likely to remain permanently. It is convenient to consider these agents under the headings: micro-organisms, including viruses, bacteria, fungi and protozoa; parasitic helminths and arthropods; and predators.

6.1.2 Micro-organisms

Micro-organisms play an important role in the control of various pests in nature. This has led many workers to investigate their application to biological control. However, trials conducted in the field have met with varied success because the infectivity and rate of spread of the micro-organisms are greatly influenced by unpredictable environmental conditions. Some parasitic micro-organisms are so host specific that even closely related pests are unaffected, while others are not particularly specific and non-target species may suffer.

However, the use of micro-organisms has various advantages. There are no special application problems since they can easily be applied in the form of dusts or sprays. Furthermore infected insects are often more susceptible to the action of conventional pesticides than uninfected ones. In spite of the difficulties associated with the cost, manufacture and storage of micro-organisms they have been used successfully in several instances.

Several bacteria have been successfully employed as biological control agents, the use of *Bacillus popilliae* and *B. lentimorbus* against the Japanese beetle *Popillia japonica* being good examples. This pest was accidentally introduced into the North Eastern states of the U.S.A. in 1916, and in the adult stage destroys a variety of trees and shrubs. The larval stage is a damaging pest of turf and commercially produced bacterial spores have been applied to infested turf where rapid control is not essential, resulting in permanent control in many of the treated areas. Infection results from ingestion of the bacillus which sporulates in the haemocoel of the insect eventually causing death of both adults and larvae. A characteristic symptom of the disease is the whitish colour of the blood, hence the name milky disease for this condition. The bacteria have been widely used in conjunction with chemicals for the suppression of the Japanese beetle in America.

Bacillus thuringiensis is pathogenic in a number of insects particularly the Lepidoptera and to a lesser degree the Coleoptera. Unfortunately the pathogen does not persist indefinitely in the field, and requires continual re-introduction. In this sense, and also because death of the pest results from the toxic effects of crystal-like proteinaceous inclusions occurring in the spores, its use has certain analogies with chemical control. This bacterium is analogous to a stomach poison, in that it must be ingested before it can affect the host.

Bacillus thuringiensis is now commercially available in Europe and America both as a wettable powder for spraying and as a dust. One of these products is marketed under the trade-name of thuricide (see Plates 8 and 9). Research indicates that dust formulations are useful where complete coverage of the underleaf surfaces of infected plants is necessary for more efficient control.

The importance of viruses in the natural control of insects is well known but it is only in comparatively recent years that they have been successfully used to control pest insects. Viruses, particularly polyhedroses and granuloses, have been successful in the control of certain major insect pests of forest trees in Canada. In 1946, a polyhedrosis virus was effectively introduced into Newfoundland for the control of the European spruce sawfly, *Diprion hercyniae*. One of the first symptoms of viral infection of the insect is a change in the colour of the abdominal segments of the larvae from green to yellow and eventually black. Infected larvae become sluggish, stop feeding and death ensues within about six days of acquiring the infection. The mode of infection is usually by feeding on contaminated food and the virus differs from most viruses in attacking the mid-gut epithelium. The European pine sawfly *Neodiprion sertifer* has also been successfully controlled in North America by using a polyhedrosis virus originally obtained from Sweden. High mortalities can be achieved by ground and aerial spraying of infested forests. The success achieved in the control of the alfalfa caterpillar *Colias philodice eurytheme*, in California using a polyhedrosis virus, indicates that viruses may be more generally useful against insect pests of field crops. This virus occurs naturally in a latent form in the pest population and can suddenly 'flare up' to produce death in large numbers of the host. One of the ways in which these outbreaks can be initiated is spraying additional quantities of virus onto the pest insects. Difficulties arise in the culture of this virus and no commercially produced sources are available at present. Attempts to use viruses against several other pests of crops have been undertaken, for example studies on the cabbage looper caterpillar *Trichoplusia ni* suggest that it is susceptible to viruses.

Myxomatosis, a virus disease of rabbits, was first introduced with dramatic results into Australia in 1950 from naturally occurring infections in wild rabbits in South America. It was introduced (whether accidental or otherwise is uncertain) into Britain in 1953 where it was responsible for reducing the rabbit population from an estimated 100 million individuals to five million in two years. Control has not persisted due to the development of resistance by the host, and the development of less virulent strains of the virus. This method of control is now considered to be inhuman and has been declared illegal in Britain (see Plate 7A).

The use of fungi in biological control programmes has been limited by difficulties encountered in their spread and survival in pest populations. Fungi are dependent upon suitable environmental conditions for their establishment and propagation in nature, one of the most important of these being a warm humid atmosphere. This has been clearly demonstrated in the case of the white muscardine fungus *Beauveria bassiana* which attacks the chinch bug *Blissus leucopterus* and also in the control of the spotted alfalfa aphid *Therioaphis maculata* in California using naturally occurring and laboratory reared fungi. Control of the apple sucker *Psylla mali* in Canada by the fungus *Entomophthora sphaerosperma* has been shown to be correlated with rainfall and this again is presumably related to humidity. The control of the chinch bug has raised particular difficulties since the normal level of disease can only be enhanced by the dissemination of extremely large numbers of spores. Clearly, considerably more research is required before an objective assessment can be made of the value of any particular fungus in pest control.

The use of protozoa as agents in biological control is theoretically an attractive proposition in view of the large number of parasitic protozoa found in insects. There are, however, difficulties in the production and storage of protozoa with the result that few attempts have been made to use them in biological control. Furthermore parasitic protozoa of insects usually cause chronic and not acute disease. Nevertheless even in chronic infestation the reproductive capacity of the host insect is often impaired, a factor of some importance in long term control. The protozoa are at least potentially useful as biological control agents. For further information refer to Steinhaus (1963).

6.1.3 Nematodes

In spite of the large number of species of nematodes parasitic in pest animals few have been utilised as control agents. Interest has been aroused by the successful use of the nematode, *Neoaplectana* sp., for the control of the codling moth (*Laspeyrisia* (=*Cydia*) *pomonella*) in America. Infection with this nematode results in the introduction of a species of bacterium into the body cavity which in turn causes a septicemia. The bacteria act as a source of food for the nematode and also produce their own antibiotic preventing the decomposition of the host by foreign bacteria. Experimental infections in the field have resulted in a 60–70% reduction in the incidence of both the codling moth and of the corn earworm, *Heliothis zea*, in treated areas. It is doubtful whether this would be sufficient to control the damage caused by these pests in some areas where numbers are large.

6.1.4 Parasitic insects

The first recorded case of a parasitic insect was reported in 1706 when the internal hymenopteran parasite, *Apanteles glomeratus*, was described from the cabbage butterfly *Pieris rapae*. This parasite was also one of the first to be used in biological control when pupae were imported from Britain in 1883 for the control of *Pieris rapae* in America. Since then parasitic insects have been extensively investigated for their application in biological control with the result that a wide variety of parasites are now in use against many pest insects in many areas of the world (see Fig. 20) (for further reference see Sweetman 1958).

Many factors influence the success or failure of parasitic insects as biological control agents. Not least of these is the ability of the parasite to seek out its host at low population densities. This searching capacity depends on the powers of locomotion of the parasite, its powers of survival and its ability to detect the presence of the host. Also the host may be physiologically unacceptable to the parasite by either not presenting the correct stimuli for oviposition or as we shall see later (pp. 145 and 167) by becoming resistant to the parasite.

Further problems arise where the ecological requirements of hosts and parasites are dissimilar. Firstly this will mean that potential host species will be isolated from potential parasites. Secondly where a host-parasite relation-

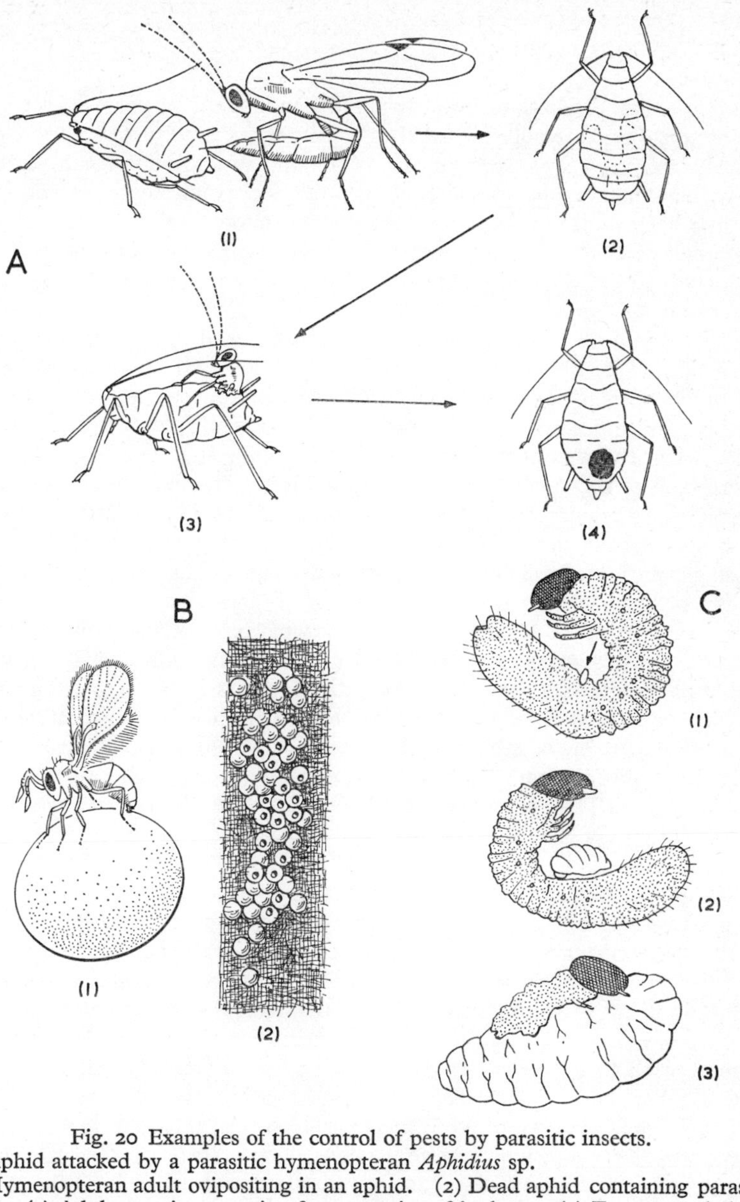

Fig. 20 Examples of the control of pests by parasitic insects.
A. Aphid attacked by a parasitic hymenopteran *Aphidius* sp.
(1) Hymenopteran adult ovipositing in an aphid. (2) Dead aphid containing parasite larva. (3) Adult parasite emerging from remains of its host. (4) Emergence hole in dead aphid.
B. (1) The parasitic hymenopteran *Trichogramma minutum* ovipositing in a moth egg. (2) Emergence holes caused by the parasite when leaving the eggs.
C. Grubs of the Japanese beetle parasitised by the hymenopteran. *Tiphia popilliavora.* (1) Egg in place. (2) Grub with parasite larva attached. (3) Fully formed parasite larva with remains of grub. Diagrams not drawn to scale.

ship does exist, the host will escape parasitism if there are areas where it can tolerate conditions unsuitable for the parasite. Parasites must also have a high reproductive capacity otherwise the host population would reach outbreak levels. Fortunately most parasites show a relatively high degree of specificity for their hosts which is clearly an advantage, in protecting non-target species, when compared with most chemical pesticides. The following examples are included to illustrate some of the factors affecting the success or failure of this type of biological control.

The tachinid fly *Ptychomyia remota* was introduced into Fiji in 1925 where it established complete control of the coconut moth *Levuana irridescens* within two years. Control was so complete that in some islands where alternative hosts were unavailable the parasite died out, and requires to be reintroduced when the pest population reaches damaging numbers. Several factors contributed to the success of this project. Firstly, the climate is suitable for the parasite; secondly the parasite has a short life cycle in relation to that of the host, enabling more than one generation of the parasite to develop in a single generation of the host; and thirdly the parasite has remarkable powers of dispersal, the winged adult being capable of dispersion up to twelve miles. This example of biological control is interesting because this particular parasite, which is indigenous to Malaya, parasitises an entirely different genus to that of the coconut moth. It is exceptionally rare in biological control for parasites to utilise a different host genus to that of the natural host in the field.

One of the factors which limits the success of parasites in biological control is the survival of the parasite population and climate is particularly important in this respect. Thus the woolly apple-aphid *Eriosoma lanigerum* which is a pest of apple trees has been successfully controlled in many parts of the world by the introduction of the hymenopterous parasite *Aphelinus mali*. However, the use of this parasite in Britain is complicated by the fact that it has difficulty in surviving the winter, and for successful control the infected hosts require to be collected and overwintered at 4–8°C. This particular example of biological control is one of the few examples of really successful pest control in Britain using parasites, another being the control of the greenhouse white fly *Trialeurodes vaporariorum* by the parasitic wasp *Encarsia formosa*. The oriental moth *Cnidocampa flavescens* which affects a variety of trees in North America has been substantially controlled using the dipterous parasite *Chaetexorista javana* imported from Japan. However, this is again an example where weather conditions interfere with control, since the parasite larvae cannot survive low winter temperatures. Similar problems occur in the control of the sugar-cane borer *Diatraea saccharalis* using the dipterous parasite *Lixophaga diatraeae*. When first introduced into Florida this parasite produced 65% mortality in the host population but adverse climatic conditions caused such high mortalities among the parasite that it now requires continual re-introduction. In spite of these difficulties numerous examples of the successful use of parasites have been reported, where climate is not a limiting factor.

PLATE 5A UNHYGIENIC CONDITIONS WHICH AID THE SPREAD OF DISEASE-CAUSING ORGANISMS
(Courtesy The World Health Organisation)

PLATE 5B PIG LITTER-MATES, THE SMALLER ONE INFECTED WITH *Metastrongylus apri*, a species of lungworm (Nematoda), THE LARGER ONE UNINFECTED

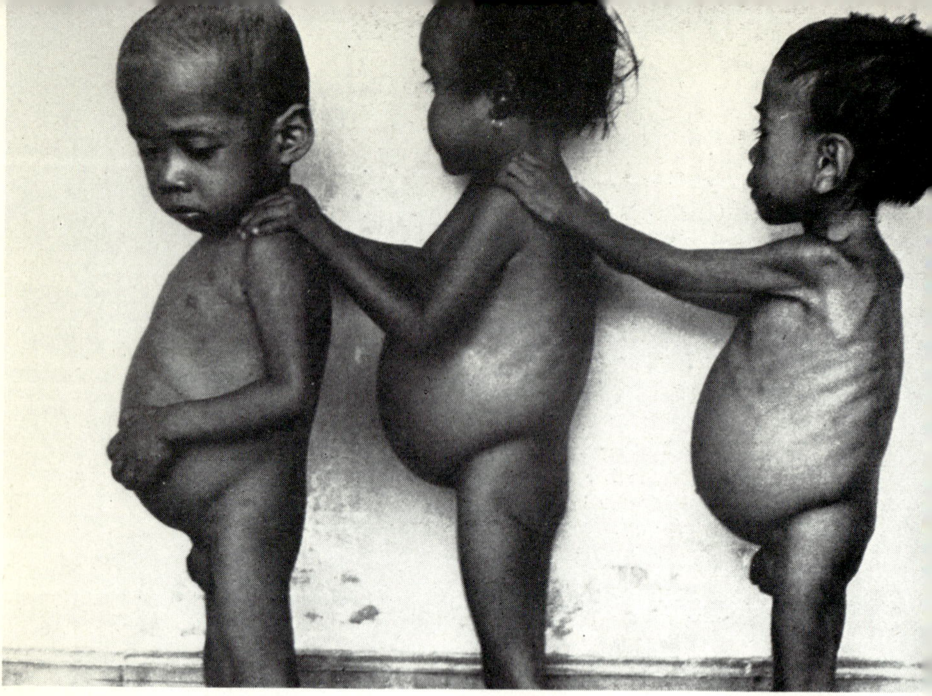

KWASHIORKOR VICTIMS. This disease condition is caused by severe protein deficiency. Parasite infestations are often found to be a contributory cause.

PLATE 6A Three year old children, showing abdominal swelling. (*Courtesy The World Health Organisation*)

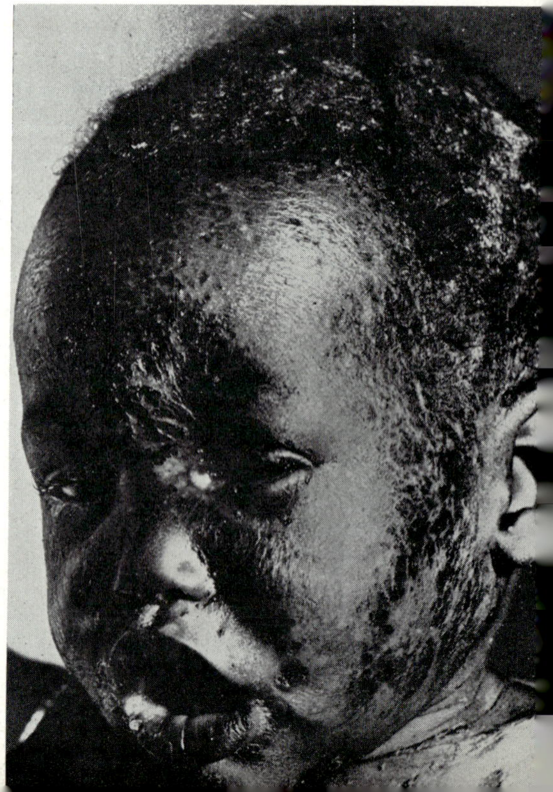

PLATE 6B Thirteen month old boy showing skin lesions. (*Courtesy The World Health Organisation*)

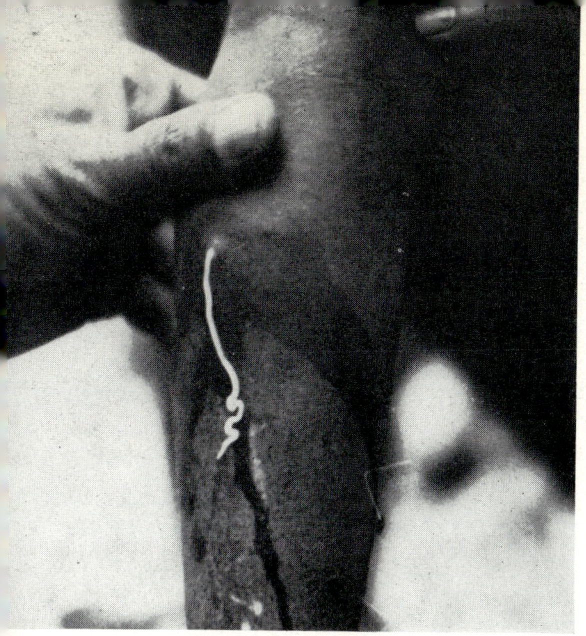

PLATE 6C GUINEA WORM (*Dracunculus medinensis*) protruding from the ulcer it causes on the human leg.
(*Courtesy of The Wellcome Museum of Medical Science*)

ATE 6D DUST APPLICATION
umans during delousing campaign
control of lice in Afghanistan.
urtesy of The World Health
ganisation)

PLATE 7A RABBIT INFECTED WITH MYXOMATOSIS

PLATE 7B INSECT INFECTED WITH VIRUS. Diseased caterpillars of the Gypsy Moth showing the typical symptoms of nuclear polyhedrosis virus. (*Courtesy C. F. Rivers*)

The larch sawfly, *Lygaeonematus erichsonii*, an important forest pest, was successfully controlled in Canada by the parasite *Mesoleius tenthredinis* originally imported from Britain, but it has now become apparent that the host is developing resistance to the parasite. This appears to result from the host's ability to phagocytise the embryo before it hatches from the egg (see also section 8.2). Although this has only occurred in the original area of introduction there are indications that resistance may spread. This is a serious phenomenon because with perhaps the exception of myxomatosis and possibly *B. thuringiensis*, this is the first case of resistance developed against biological control agents.

6.1.5 Predators

The use of predaceous insects for the control of pest organisms has been practised for a considerable period of time. Since ancient times it has been common practice for Chinese citrus growers to place nests of the predaceous ant, *Oecophylla smaragdina* in mandarin orange trees to reduce the number of foliage feeding insects. Similarly the date growers in the Yemen have traditionally used predatory ants to control insects pests on date palms. The beneficial nature of ladybirds which destroy aphid colonies has equally been known for centuries.

The first highly successful case of biological control occurred in 1888. In the previous year the citrus growing areas of California were plagued by massive infestations of the cottony cushion scale *Icerya purchasi* which had been accidentally introduced from Australia several years earlier on acacias. In 1888 the vedalia beetle *Rodolia cardinalis*, together with the parasitic fly *Cryptochaetum iceryae*, were introduced from Australia into the citrus growing areas of California. The project was a complete success, both insects having contributed to this, but the vedalia beetle playing the major role. This exercise established biological control as an economic method of reducing pest infestations. Since this time many biological control campaigns using predators have been attempted, many with a considerable degree of success, and the following examples are included to illustrate the range of predatory animals used.

A number of scale insects are successfully controlled by predatory insects. The eucalyptus scale *Eriococcus coriaceus* has been controlled in New Zealand since 1905 by the coccinellid beetle *Rhizobius ventralis*. Four species of harmful scale insects (*Ischnaspis longirostris*, *Chrysomphalus ficus*, *Eucalymnatus tessellatus* and *Pinnaspis buxi*), all of which attack coconuts in the Seychelles, are controlled by four species of coccinellids introduced from East Africa in 1939. The method is so successful that additional methods are unnecessary.

The sugar cane aphid *Aphis sacchari* has been a pest in Hawaii since 1896. The lady beetle, *Coelophora inaequalis*, is a remarkably successful predator of this pest. A list of other insects successfully controlled by predatory insects can be found in De Bach (1964) and Sweetman (1958).

Predatory mites are also important agents for use against pest animals. Populations of the springtail, *Sminthurus viridis*, a pest of pastureland in Australia, have been effectively reduced over large areas since the introduction of the mite *Biscirus lapidarius*. Predatory mites have also been used against harmful plant-feeding mites, and one of these predators *Phytoseiulus persimilis* is proving so successful in field trials in glasshouses in Great Britain that the use of acaricides to control glasshouse red spider mite (see Plate 1B) may well ultimately become unnecessary. In a similar way predaceous snails have been employed against phytophagous snails in some of the Pacific islands.

Various species of fish, particularly belonging to the families Cyprinodontidae and Poeciliidae have appreciably reduced the population of mosquito larvae in tropical and sub-tropical regions. However, most of this success has been achieved in small areas of water, particularly artificial pools. In larger areas biological control requires to be supplemented by insecticides before effective results are obtained. Other vertebrates, for example the toad *Bufo marinus*, have been used as agents in biological control. This toad produced effective reduction in the numbers of rhinocerus beetle, *Strategas barbigerus*, of sugar cane, for twenty years after its introduction into Puerto Rico in 1920. Unfavourable environmental conditions such as poor food supply and lack of breeding sites have caused a reduction in the number of toads and consequently their effectiveness.

The most recent and perhaps outstanding success in the use of mammals in biological control has been achieved in Newfoundland where the masked shrew (*Sorex cinereus cinereus*) has been introduced to augment the natural enemies of the larch sawfly (*Lygaeonematus erichsonii*). In 1958 twenty-two shrews were introduced from the mainland of Canada and by 1964 the numbers per acre exceeded those in the original area on the mainland. This has resulted in a marked decline in the incidence of the larch sawfly in Newfoundland due principally to the predatory habits of this insectivore against the soil dwelling cocoons.

6.1.6 The relative value of predators and parasitic insects in the biological control of arthropods

Predators have been described by De Bach (1964), with particular reference to arthropods as "organisms that consume more than one individual host (prey) in order to reach maturity. On the other hand parasites develop to the adult stage on a single host individual". The fact that predators consume more than one individual host may lead one to assume that the predator is the more effective control agent. There is, however, some indication that this is not always the case. Clausen (1958) reports that of 95 species of insects introduced into the U.S.A. for control purposes 81 were parasites and only 14 were predators. Furthermore the majority of the successful cases of biological control are attributed to the use of parasitic insects.

One of the suggested reasons for this success is that since the parasite requires only one host individual it can operate successfully at very low host population densities. In contrast, predators require high pest densities before they can successfully complete their life cycles. In addition the searching capacity of parasites, already mentioned (see section 6.1.4) is highly adapted and often more efficient than that of predators.

The effectiveness of predators in control should not be underestimated however. Their importance is illustrated when indiscriminate use of pesticides reduces the predator population, resulting in the alteration of the status of the prey species to that of a significant pest. Some of the most successful and dramatic examples of biological control have resulted from the use of predators, for example the use of the vedalia beetle against the cottony cushion scale in California, the reduction of coconut scale (*Aspidiotus destructor*) in Fiji by the beetle, *Cryptognatha nodiceps*, and the sugar cane hopper (*Perkinsiella saccharicida*) in Hawaii by the predatory bug *Cyrtorhinus mundulus*.

The success of these particular predators has been attributed to their specific feeding habits which enable them to respond very quickly to increases in the pest population. The ability of a predator to exploit alternative prey on the other hand is also advantageous, in that it enables it to survive if the density of the pest species is low. The role of vertebrate predators in the control of insect pests is not clear. Indeed, where the diet of the predator is varied the populations of most vertebrate predators of insects are little, if at all, affected by increases in the density of a particular pest species. Thus, for example, bird populations generally show poor numerical response to the presence of the pest (by increasing in numbers proportional to those of the pest) and they are therefore unlikely in the long term to have any profound effect on the population density of a pest. At epidemic levels any predatory effect is often swamped by the large number of pest individuals. Thus we see that the winter moth (*Opherophthera brumata*) in Britain is little influenced by the predatory effects of tits. However, while this may be true as a generalisation there are special situations where vertebrates may be important, for example where the predator is highly selective and hence takes larger numbers of the pest during epidemics. Thus a review of the role of vertebrates in the control of insect pests of Canadian forests has described birds as having an important effect on the incidence of the spruce budworm *Choristoneura fumiferana* since they feed highly selectively on pupae and larvae.

6.1.7 *Evaluation of biological control*

It is not easy to find acceptable criteria with which to assess the value of biological control over other methods, but it is sometimes possible when comparisons are available between the cost of a biological and of a chemical control method. However, such information is rarely available, often because the gains of biological control are not immediately apparent, perhaps taking many years before they are fully felt. Furthermore, in view of the problems of resistance to pesticides and also the toxic hazards resulting from the use of some chemicals (see section 6.1), it might be more acceptable in the long term to choose a biological control method, albeit less immediately effective, than a chemical one. Since successful biological control is permanent or semi-permanent one can assume that the cost/potential benefit ratio (see section 1.2 and section 9.2) will be high compared with chemical methods and will grow for each year that successful control is maintained. Many cases of apparently successful biological control have not withstood the test of time while others, although excellent under experimental conditions, have failed to work in the field sufficiently well to merit their application. Added to this there is the difficulty of culturing, marketing and distributing many biological control agents, often making the technique economically unfeasible.

In spite of these difficulties numerous instances of successful biological control are available. Sweetman (1958) lists 80 highly successful examples and De Bach (1964) adds extensively to this list. A wide variety of pests have been controlled although the vast majority of these are insects. Many of these pests are relatively sedentary or are at least gregarious, which makes them more vulnerable to parasite and predator attack. About 40% of the successes have been against coccids (soft scales, armoured scales, and mealybugs) and the remainder against other Homoptera, Lepidoptera and Coleoptera. It has been suggested that the high proportion of coccids controlled results from their occurrence on expensive crops and the fact that they are easily transported into other countries. Furthermore the early successes against particular coccids coupled with the difficulty of using chemical methods of control against this group, have stimulated research into the control of other coccid pests.

The majority of beneficial predators and parasites are insects, but mites, molluscs and vertebrates are also included. The Hymenoptera contains the largest number of beneficial parasites while the Diptera, bacteria and viruses are also important. Among the cases of successful biological control, parasites are seen to provide four times as many examples as predators.

There is considerable controversy regarding the future of biological control. Some authorities consider that the point of diminishing returns has now been reached and that the most worthwhile achievements in this field have already been made. Others consider that there is still considerable scope for extending this method of pest control. Other critics consider that this method is most likely to be successful on tropical islands since many of the successful cases have occurred in Fiji and Hawaii. This has been attributed to climatic conditions, often very favourable to the enemies of pests, and furthermore, the large number of pests of foreign origin on such islands makes the importation of natural enemies more likely to produce startling results. However, when a careful analysis is made of the occurrence of successful cases of biological control it can be seen that 60% have been achieved not on islands but on continents. Furthermore, considerable success has occurred in temperate regions particularly in America.

With regard to the future for biological control, leading authorities consider that even in California, where many successes have been achieved, there is still scope for the importation of natural enemies. Thus although citrus pests have been the subject of intensive study in this respect for seventy years promising new natural enemies are still forthcoming. Furthermore of the ten most serious pests of crops in California only two (the California red scale *Aonidiella aurantii* and the spotted alfalfa aphid) have been extensively investigated as subjects for biological control.

It seems likely that biological control will become increasingly important in the future, particularly in view of the increasing travel between countries which is likely to result in a greater number of introduced pests, and also because of the problems of resistance to chemical pesticides. However, in view of the problems still to be overcome, the immediate future for biological control probably lies in its closer integration with cultural and chemical methods.

6.2 Direct Reduction Using the Pest Species Itself

The use of attenuated individuals of a pest species to evoke immunity in animals is an important aspect of pest control, and this is dealt with in section 2.3. We are now concerned with the release of sterile individuals into natural populations of pests, and also with the application of chemosterilants to pest populations in the field. These sterile individuals compete with normal ones for mates, with the result that there is a drastic reduction in numbers in the following generation, since many of the eggs fail to hatch. This method is most effective where the female mates only once in the breeding season and where there is no difference between sterile and normal insects in their ability to obtain mates. This means that sterilisation must not reduce the competitive vigour of the pest by, for example, altering its behaviour patterns, survival times, etc. Knipling (1962) has shown that in a theoretical model (refer to Table 12) sterilisation produces much faster reduction in the size of a pest population than does the use of conventional insecticides. In Table 12, assuming the pest population increases at a fivefold rate in each generation, and an insecticide produces 90% mortality in each generation, then the population will be eliminated within twenty generations. The application of a chemosterilant producing 90% sterilisation to each generation results in complete elimination of the pest in five generations. The reason for this is that in the case of the insecticide 100 000 of the parental generation survive and at a fivefold rate of increase they will produce 500 000 offspring. On the other hand the sterilant leaves only 100 000 of the parents capable of mating and their chances of meeting a mate in a 90% sterile population are one in ten. Therefore, only 10 000 matings are fruitful and with a fivefold rate of increase 50 000 offspring result. As Table 12 shows, this process is repeated in subsequent generations. Sterilisation is thus doubly effective in reducing the size of the pest population. Whether this advantage is likely to occur in practice requires to be investigated.

Sterilisation can be achieved either chemically or by irradiation. The eradication of the screw-worm fly *Cochliomyia* (=Callitroga) *hominivorax*, from the island of Curaçao in the Dutch West Indies has been accomplished by the release of male flies sterilised by X or γ irradiation. This pest is a destructive parasite of livestock, laying its eggs in the wounds of warm blooded animals. The larvae hatching from the eggs feed on the flesh for about five days, after which they fall to the ground to pupate. Several weeks later the adult flies emerge from the pupae and begin to lay eggs on fresh hosts. In heavy infestations several thousand larvae may parasitise a single animal and death commonly results. All domestic animals, many wild mammals and occasionally man are attacked and it has been estimated that the average loss to the livestock industry of the United States due to this pest was of the order of 25 million dollars annually.

In the early 1950s the United States Department of Agriculture instituted a research programme to investigate the feasibility of releasing large numbers of sterilised male flies into screw-worm infested areas. These were reared in the laboratory and sterilised with X or γ irradiation. Mating of sterile males and fertile females produced sterile eggs, the females continuing to produce sterile eggs for the remainder of their breeding life (three weeks). Early field trials on an island off the coast of Florida successfully reduced the population of screw-worm flies to a very low level but complete eradication was not achieved, possibly due to migrations into the area from nearby infested islands, and to obviate this effect field trials were conducted in 1954 on the island of Curaçao which is 65 kilometres from the nearest land mass. Sterile males were released at the rate of 160/square kilometre/week (natural population less than 40/square kilometre) and the degree of fertility of eggs collected from penned control animals estimated (see Table 13). The results of this experiment showed a population decline which closely followed the prediction made by means of a theoretical model similar to that given in Table 14.

By 1957 no further cases of screw-worm fly had been recorded in Curaçao and it was concluded that eradication was complete. In 1958 a control programme was initiated in the south-eastern states of America and within a year almost complete eradication was achieved. Two billion sterile males and females were released in eighteen months into an area of 180 000 square kilometres which included Florida and parts of Georgia and Alabama.

In view of the success achieved against the screw-worm fly, efforts are being made to extend this type of control to other pests. The oriental fruit fly, *Dacus dorsalis*, on the Pacific island of Rota, was not controlled by the released sterile flies. However, the release of 17 million irradiated flies on the adjacent island of Guam, where the natural population of the pest had been lowered by typhoons, resulted in complete eradication. The melon-fly, *Dacus cucur-*

bitae, has also been eradicated from Rota by the release of large numbers of sterile individuals. Research into the control of other pests using individuals sterilised by radiation or chemicals is being carried out at present.

One of the major problems associated with the use of irradiated insects is the cost and labour involved in rearing sufficient numbers of the pest species for release into the infested area. The resident population of the pest organism should not be too large to be 'swamped' by the number of sterilised males which can be released. Another drawback is that release of large numbers of pests may cause a considerable, albeit temporary, increase in damage. This would not apply where the pest is harmful only in its larval stages as in the screw-worm fly, but would be so for most pests causing damage as adults. Chemical sterilisation in the field might overcome many of these problems.

Certain chemicals are capable of producing irreversible sterility in pest insects without adversely affecting the mating behaviour and the length of reproductive life. Many of these chemicals belong to classes of compounds known to inhibit the development of cancer tumours. An approved method of classifying chemosterilants has not hitherto been devised and it is common to use the system devised for antitumour compounds. Two of the important groups of compounds included in this system are alkylating agents and antimetabolites. The former group includes the aziridines (formula XLI, section 4.2.7), many of which are important chemosterilants in insect control. The mode of action of these aziridines in the production of sterility is not yet clear. The latter group, the antimetabolites interfere in some way not at present understood, with essential metabolic processes and the chemosterilants included in this group have been shown to interfere with the synthesis of nucleic acids.

There are still many problems associated with the chemosterilants available at present, for example it is important for the insect to remain at least as competitive as the normal insects in the population. In trials on the housefly there is some indication that sterilised males are in fact more successful in mating (as measured by the ratio of sterile to fertile eggs laid). How this occurs is open to conjecture but one possible explanation may be that mating by sterile males nullifies in some way previous successful and fertile matings. Whether such results are applicable to conditions in the field remains to be seen. Similar experiments on mosquitoes failed to show any increase in competitive vigour.

One advantage of chemosterilants is that by altering dosages, specificity may be obtained, thus the stable fly *Stomoxys calcitrans* requires 45 times less apholate for complete sterilisation than does the screw-worm fly. Field trials conducted on houseflies on a small island off the coast of Florida reduced the hatch of eggs to 1% in five weeks. This indicates that chemosterilants are at least potentially useful control agents but the success achieved in small isolated areas may be difficult to reproduce in areas where invasions from outside the control area are possible.

There are of course drawbacks to the use of sterilisation techniques. Since there is no immediate reduction in the pest population this method of control is not likely to appeal to the individual farmer who, for example, is faced with a destructive pest outbreak on a growing crop. It might be possible to overcome this objection if a chemosterilant was incorporated with a conventional insecticide thereby sterilising any insects surviving the insecticide treatment. In the case of widely dispersed pests the chemosterilant would need to be distributed over wide areas to be effective and this would be expensive and require co-operative action by growers or the authorities. Chemosterilants of much lower toxicity to non-pest organisms than those already available are required and further research into possible side effects is required before widespread application in the field could be envisaged. These problems might be more easily met than similar problems with more conventional insecticides since chemosterilants tend to be more specific. It has been suggested that this specificity could be increased still further if the chemosterilant was incorporated with a specific attractant.

There is no indication at present that insects develop resistance to chemosterilants but it cannot be assumed that this will not occur in the future. Resistance could develop if individuals in a pest population were subjected to sub-effective doses of chemosterilants. Any individuals showing a higher tolerance to these chemicals would survive and breed successfully. Selection for this resistance in the population subjected to chemosterilant treatment would then be likely.

TABLE 12 THE THEORETICAL BASIS FOR POPULATION REDUCTION BY STERILISATION*

Relative reduction of a theoretical population of a pest subjected to:

(1) No control
(2) 90% kill with insecticides each generation
(3) 90% sterilisation of both sexes in each generation

	Number of insects resulting from		
Generation	*No control at assumed 5× increased rate*	*Insecticide giving 90% kill applied each generation*	*Treatment giving 90% sterilisation applied each generation*
Parent	1 000 000	1 000 000	1 000 000
F 1	5×10^6	500 000	50 000
2	$5^2 \times 10^6$	250 000	2500
3	$5^3 \times 10^6$	125 000	125
4	$5^4 \times 10^6$	62 500	6
5	$5^5 \times 10^6$	31 250	0
10	$5^{10} \times 10^6$	976	—
15	$5^{15} \times 10^6$	31	—
20	$5^{20} \times 10^6$	1	—

* Modified from Knipling, E. J. (1962) *Journal of Economic Entomology* **55**, 782–786.

Sterilisation presents an attractive possibility as a control measure. Irradiation-produced sterility has been shown to produce effective control in some insects, but the breeding and release of irradiated pests is likely to remain prohibitively expensive except in the case of very destructive pests. Furthermore the results produced in the case of the screw-worm fly may not be so easy to reproduce in pests mating many times in the breeding season and in those in which the adult is the damaging stage. Chemosterilants offer more attractive possibilities particularly if specificity and lack of side effects can be assured, and the future here may lie with the development of specific attractants. While chemosterilants have not reached the practical stage of development where compounds are available for safe use, the future holds exciting prospects for sterilisation as a method of pest control.

TABLE 13 RESULTS OBTAINED AFTER INTRODUCING 154 STERILE MALES OF SCREW WORM FLY/SQUARE KILOMETRE/WEEK INTO THE ISLAND OF CURAÇAO AND SAMPLING PENNED ANIMALS FOR THE DEGREE OF FERTILITY IN THE EGGS LAID*

Weeks after start of experiment	*Number of egg masses*		*% sterility of egg masses*
	Fertile	*Sterile*	
1	15	34	69
2	17	38	69
3	17	36	68
4	10	37	79
5	7	42	86
6	3	23	88
7	0	10	100
8	0	12	100
9	0	0	—

* Knipling, E. J. (1960) *American Journal for the Advancement of Science*, 169–182.

TABLE 14 THEORETICAL POPULATION DECLINE IN EACH SUBSEQUENT GENERATION WHEN A CONSTANT NUMBER OF STERILE MALES ARE RELEASED AMONG A NATURAL POPULATION OF ONE MILLION FEMALES*

	Generation			
	1	*2*	*3*	*4*
Number of unmated females in area	1 000 000	333 333	47 619	1107
Number of sexually sterile males released	2 000 000	2 000 000	2 000 000	2 000 000
Ratio of sterile to fertile males competing for each female	2:1	6:1	42:1	1807:1
Assumed percentage of females mated to sterile males	66·7	85·7	97·7	99·95
Theoretical population decline of fertile females	333 333	47 619	1107	less than 1

* From Knipling, E. J. (1960) *Op. cit.*

Chapter Seven
INTEGRATED REDUCTION

Chapters 2, 3 and 4 of this book have given some indication of the diversity of methods available for the control of pest animals. A great deal of money is spent on fundamental research and on the practical application of these methods to pest populations. Many spectacular successes have been achieved, and the success of any one method in dealing with a particular pest problem has often tended to produce enthusiastic supporters, stressing the advantages of this particular method of control and emphasising the disadvantages of all others. This is particularly true of applied biological control, which is sometimes advocated as a natural control method and therefore endowed with some innate superiority over chemical control methods for all pest problems.

However, the applied zoologist, faced with a practical problem, must call upon whichever of the weapons in the control armoury are likely to produce the maximum economic benefit combined with the least risk of damage to the environment. At the present time chemical control methods are those which are most extensively used, and upon which most resources are expended. They are also those methods which are subjected to the severest criticism.

There is a two-fold basis for this criticism, in the first place the hazards associated with the use of chemicals give rise to concern and in the second place pesticides are frequently indicted as inadequate or even ineffective agents of successful control. The hazards arising from the use of chemicals are discussed in Chapter 8, together with the failures in chemical control operations due to the development of resistance. However, there are other grounds upon which the effectiveness of chemical control is often questioned and these must now be considered.

The ways in which various factors influence the numbers of an animal species have been discussed in Chapter 1. All animal communities contain both abundant species, and less common, or even rare species. If an abundant species causes economic damage to one of the resources valued by man it is considered a pest. We have described the way in which all animal populations show fluctuations in numbers. If the population of the abundant species shows fluctuations which are always above the population level causing economic damage, then this species is a permanent pest. On the other hand, if only the peaks of the fluctuations rise above the minimum level for economic damage, the species causes periodic outbreaks of economic damage. These fluctuations in numbers result from the variations in the pressure various environmental factors impose on a species. This natural regulation of population size is discussed in section 1.4.

The argument is put forward that a pest is an animal which is too abundant when seen through human eyes; this excessive abundance is due to low environmental pressure, conditions being basically favourable to the species. The use of chemical pesticides does not permanently affect the environment to the detriment of the pest. It is therefore argued that after a period of chemical slaughter, the surviving members of the species will multiply, possibly at a greater rate than formerly, because of the absence of competition with others of their own species, with the result that the pest problem will recur. Even if the chemical succeeds in exterminating the species in a given area, recolonisation from surrounding sources of infestation may occur rapidly. On the basis of reasoning such as this, critics of the chemical control techniques often describe them as 'expensive flyswatters', producing a sudden, but only temporary check in the population.

In practical terms, of course, the 'expensive flyswatter' may be all that is required. For example an insecticide may provide temporary protection for a valuable crop just before harvesting, or an anthelmintic may protect a young animal from heavy parasite attack while it grows through a particularly susceptible stage. In these cases, of course, the suitability of the environment for the pest is undergoing a change and the chemical is successful simply as a means of 'buying time' for the farmer.

An even more serious criticism of chemical treatment is that the ultimate result may be that the pest population recovers to the point where it fluctuates about an even higher mean density than before treatment. This may occur if a major part of the environmental pressure acting upon the pest consists of the attack of parasites or predators. In the case of insect problems, unless a highly selective insecticide is used the insect parasites and predators will suffer as much as the pest on which they feed. Indeed they may be even more susceptible to the chemical than is the pest, in which case the long term effect of treatment is to alleviate the environmental pressure on the pest population, allowing it to reach even higher densities.

There is yet another possibility to be considered when the ecological consequences of chemical control are being examined. An organism may be present in a community in only low numbers, and this low population density may be primarily due to the activity of parasites, predators or pathogens. If this organism feeds upon some product of value to man it is a potential pest, even though its normal density is so low that the practical agriculturalist is completely unaware of its existence. Pesticides may then be introduced into the environment in order to kill some quite unrelated species of organism, and if, as is frequently the case, this chemical treatment involves the use of a wide-spectrum agent, predators and parasites of the rare organism may be affected more than the organism itself, with the result that its numbers increase, possibly to the point where it acquires major pest status. As long as the chemical treatment is repeated, the depredations of the new pest will continue. The unsuspected biological control has been broken down and the hitherto rare animal becomes exceedingly abundant.

This possibility of the creation of new pests is one of the most serious criticisms that can be levelled at chemical control procedures, and has already been briefly discussed in section 1.3.3. Here mention was made of the now classic case of the fruit tree red spider mite *Panonychus* (=*Metatetranychus*) *ulmi* which was a relatively uncommon arachnid in orchards in south-east England until 1922. The tar oil washes introduced as dormant sprays for apple trees in 1923 had no effect on the mite eggs, but destroyed predaceous mites, bugs, and beetles such as ladybirds. The result was a steady growth in red spider populations and the emergence of a new pest. This process was speeded up by the introduction of DDT after the 1939–45 war. This insecticide kills many of the predators and parasites of the red spider, but it is not an acaricide. Other changes in chemical treatments also acted in the mites' favour. New and more effective fungicides were introduced into orchards and these also had an indirect effect on mite populations owing to their toxicity to the enemies of *P. ulmi*. The decrease in the environmental pressure on the mite produced a major pest, requiring a new range of chemicals to be developed for its control.

Fungicide treatments of crops may have a more direct effect on a pest population, since part of the environmental pressure acting on the pest may be due to fungus disease. This is true of the apple sucker *Psylla mali*, which is often destroyed by a disease caused by the fungus *Entomophthora sphaerosperma*. Observations have been published which show that the post-war change from inorganic to organic fungicides, for the improved control of apple fungal diseases, also resulted in destruction of this pathogenic fungus. In at least one case the expansion of *Psylla mali* populations to a level where they produced serious economic damage has been attributed to this cause. It is obvious that the effects of chemicals on biological control agents should be considered before introducing chemicals into the environment of a community. Wherever possible application techniques must avoid the destruction of useful parasites, predators and pathogens. By careful choice of pesticide, formulation, and timing of application, the benefits of both biological and chemical control agents can be obtained, rather than one acting antagonistically to the other. This type of approach has been termed Integrated Control, which might be defined as the use of chemicals to reduce pest damage, without disastrously reducing the capability of the biotic environment of the pest to adversely influence pest numbers. However, the term integrated control is also often used in a much wider sense these days, to cover the use of any combination of control measures on the basis of ecological information about the pest and its environment.

Nevertheless, for practical purposes, at the present time, the term integrated control is applied primarily to the integration of chemical and biological techniques. The value of this approach obviously varies according to the importance of other organisms in the pests' environment. In some circumstances they may not be at all important, as for example in the case of certain insect infestations of stored products, where the consequences of chemical application may be easily predicted and there is little need to worry about the use of highly selective chemicals or formulations.

As the community in the treated environment becomes more complicated, the repercussions arising from treatment become more difficult to predict. The treatment of forest or orchard trees, for example, demands an integrated approach if long-term economic success is to be obtained. The indiscriminate use of wide-spectrum chemicals on particular pests in orchards may produce changes in the community in the ways outlined above. New pests would then appear requiring additional treatment perhaps with other pesticides. Eventually a situation arises in which pest damage is kept to a minimum only as a result of a programme of spraying which involves the very frequent application of a considerable diversity of chemicals. This becomes the only method of maintaining the highly artificial community which has been created.

The more rational approach to the use of pesticides in orchards was pioneered by A. D. Pickett and his co-workers in Canada. By detailed investigation into the ecological effects of chemicals they developed an integrated approach to the control problems in their orchards. These pioneer investigations have subsequently been amplified by other workers on various crops in many parts of the world so that gradually pesticides are beginning to serve man as 'a rapier and not as a bludgeon'.

In practical terms, the preservation of biological control agents may be enhanced by simple changes in formulation, application rates or frequencies. In addition, it may be necessary to select a narrow spectrum or specific pesticide in place of a wide spectrum compound. In the long term a less toxic, but more specific chemical may produce better control than a highly toxic broad spectrum compound. Thus if 60% of a pest population is killed and none of its parasites or predators, the ultimate level of damage may be lower than if 95% of the pest population is killed and all of its predators and parasites.

Some of these considerations are well illustrated by a table showing the damage caused by three major pests in Canadian orchards which was published by Pickett (see Table 15).

Lead arsenate had been used for codling moth control from 1930 to 1949 and then the growers switched to the more effective wide spectrum insecticide DDT. This was used until 1954 and its wide spectrum effect resulted in a reduction in pollinating insects, as reflected in the low numbers of solitary bees over this period. DDT is also highly toxic to most predaceous and parasitic insects, with the result that pest numbers of other species of cater-

TABLE 15 THE EFFECT OF AN INTEGRATED CONTROL PROGRAMME ON CERTAIN INSECTS IN CANADIAN ORCHARDS*

	AVERAGES FOR 80 ORCHARDS				
	Injuries per 100 apples				
Year	*Codling moth*	*Eye-spotted bud moth (Spilonota ocellana)*	*Fruit worms*	*Percentage free from insect damage*	*Solitary bees on apple blossoms*
1948	30·8	4·0	1·1	73·1	No data
1949	8·1	5·4	1·9	76·3	No data
1950	11·2	7·3	1·8	75·0	No data
1951	13·9	6·4	2·8	81·1	8·5
1952	13·4	3·1	3·6	76·0	4·9
1953	14·9	0·5	3·5	79·0	1·7
1954	5·5	0·1	2·0	91·1	2·7
1955	6·5	0·02	1·4	91·8	4·8
1956	6·9	0·03	0·9	90·4	5·8
1957	3·3	0·08	1·2	92·1	10·6
1958	4·1	0·6	0·4	93·3	16·6
1959	2·6	1·4	0·05	93·9	19·0
1960	2·2	1·6	0·03	93·2	11·5

* After Pickett (1962) *World Review of Pest Control*, **1** (2).

pillars, together with mites and scale insects arose from the use of this chemical. In 1954 DDT was replaced by ryania as a control agent for codling moth. Although ryania is less toxic to the moth, the amount of moth damage dropped abruptly from 1954 onwards. This is attributable to the fact that ryania has very little effect on parasitic and predaceous insects and the effect of the chemical on the pest was thus substantially reinforced by biological control agents. Apart from the increase in the amount of undamaged fruit the actual cost of control was reduced in that the number of applications of insecticides found to be necessary was drastically reduced.

The overall increase in efficiency which results from the successful blending of two or more control techniques, is not confined to insect pest problems. The term 'integrated control' is seldom applied to the destruction of worm parasites, and yet exactly the same outlook may be involved. For example, the presence of worm larvae in a host may invoke the acquisition of a protective immunity, and the use of a larvicidal anthelmintic may prevent this immunity developing. On the other hand, the presence of adult worms often prevents larvae which are present from completing their development, and the use of an anthelmintic which destroys the adults can be followed by a sudden and unexpected deterioration in the host's condition as the larvae are then able to mature (see also 8.2.2). The correct use of larvicidal compounds, or compounds acting only against adults, therefore necessitates a consideration of the host animals immunological status and level of debilitation, and an attempt must be made to integrate the biological reaction of the host with the toxic activity of the chemical.

Successful integrated control involves the application of pesticide technology in the light of information provided by fundamental physiological and ecological research. If the natural and artificial environmental pressures acting upon a pest can be manipulated to act towards a common goal, the word control can then be used in a meaningful sense.

Chapter Eight
THE SIDE EFFECTS OF PEST CONTROL

8.1 Effects on the Environment and Non-pest Organisms

8.1.1 Introduction

Few methods of pest control produce no side effects on the environment, and the numerical reduction of one member of a community may have far reaching effects on its other members. If the pest organism occupies a vital position in a food chain any alteration in its numbers may markedly influence the populations of animals dependent upon it as a source of food, and also of those animals on which it feeds. Alternatively the extermination of animals acting as reservoir hosts for parasites may result in alterations in the surrounding fauna. Rabbit control in Britain provides a good example of how the flora may also alter when the pressure of a pest population is suddenly reduced (section 8.1.3). The indiscriminate use of control measures may also adversely affect useful plants and animals either directly, or by indirect influences similar to those outlined above. Some of these ecological implications of pest control are of considerable importance and have already received attention (Chapter 7).

Pest control may produce side effects having long lasting consequences, such as the sociological changes resulting from the eradication of disease. Control of the mosquito and tsetse fly in certain areas of the world has made possible the exploitation of large areas of land for agricultural and industrial development. In a similar way the economic and health status of whole communities may be improved by the utilisation of appropriate control measures.

Side effects may also be of a less beneficial type, such as deaths of non-pest animals and the contamination of consumers, and it is on these direct consequences of man's use of various control measures that the emphasis will here be placed. In this context perhaps the most obvious and important of these side effects are those resulting from the use of chemical control, the widespread use of chemical pesticides having focused attention on the possible hazards they represent to animals, including man.

8.1.2 Side effects of pesticides on man

While considerable care is taken to select pesticides which have as low a toxicity as possible to non-target organisms, many pesticides in use are toxic in varying degrees to a wide range of animals. The effect of a toxic pesticide on man or his domestic animals may be acute, where a single dose causes death or illness, or chronic, where effects are produced by the accumulation of sub-lethal doses over an extended period of time. Such chronic effects should not be overlooked because it is not often possible to forecast what they may be.

The toxic hazards to man arise firstly from the manufacture and application of pesticides. The acute toxic hazards on British farms do not seem to be very great since between 1956 and 1960, an average of 20 000 non-fatal accidents due to all causes occurred each year, and a further average of 140 accidents per year were fatal; nevertheless, not a single case of death due to the handling of toxic chemicals was recorded during this period, and an annual average of less than five of the non-fatal cases could be attributed to the faulty handling of such chemicals. In spite of the precautionary measures taken, concern has arisen regarding some of the persistent organochlorine compounds which are generally characterised by their low specificity and extreme stability. These properties have focused attention on the possible indirect harmful effects to the public of the residues of pesticides in soil and food.

The 1964 Advisory Committee, set up by the United Kingdom government to review the problems resulting from the use of toxic chemicals in Britain, investigated the presence of residues of DDT, BHC, heptachlor epoxide and dieldrin in butter, milk, mutton kidney fat, beef fat and potatoes. The results showed that these pesticides (with the exception of heptachlor epoxide which was only present in measurable quantities in milk) were present in each group of foodstuffs investigated. In the case of mutton kidney fat the high dieldrin levels found were attributed to the fact that at that time, sheep dipping with this chemical prior to slaughter was practiced. This may have represented a source of the dieldrin residues found in human fat. The aldrinated fertilisers then used for potatoes were also a possible source of residues found in man. Precautionary measures were therefore suggested for these compounds and are discussed on page 165.

The question which immediately presents itself, having demonstrated that residues do occur in human fat, is the possible hazards this situation represents. Considerable publicity has been given to statements in Rachel Carson's 'Silent Spring' which imply that the organochlorine insecticides are powerful liver poisons, but it has subsequently been concluded that with the exception of carbon tetrachloride these fears are not justified. Experiments with DDT, aldrin and dieldrin have led some investigators to describe these compounds as a weak carcinogenic hazard. The results were by no means conclusive and were treated with reservations by the investigators themselves. The situation is summed up in the words of the 1964 Advisory Committee; "While it might be considered a sound policy to treat everything capable of producing tumours of any type under any experimental conditions in any strain of animal as a potential hazard to man, too strict an interpretation of such a policy could lead, for example, to the exclusion of glucose or salt from the diet". With regard to the levels of DDT and dieldrin accumulating in human fat there is no evidence at present that they have any harmful effects. Much larger levels of certain of the persistent pesticides administered to volunteer patients have produced no observable adverse effect. However, little information is available on possible long term effects, and more experimental evidence is required on this aspect of pesticides before an objective assessment can be made.

8.1.3 Side effects on domesticated animals and crops

Domesticated animals probably have levels of residues in their bodies at least as high as those found in man and it is probable that farmstock have even higher levels in view of the likelihood of their consuming herbage, other sprayed crops and dead pests. The problem of long-term effects may not be so acute as in humans, however, since the life span of domesticated animals is generally shorter. More direct deleterious effects may be seen in the treatment of parasitic diseases, and such effects may result purely from the misuse of a particular control measure. Fatalities may result for example in the treatment of cattle with carbon tetrachloride for control of the liver fluke *Fasciola hepatica*, and many of the other anthelmintics in use at present may produce varying toxicological side effects of varying severity. This may be tolerated in the case of domesticated animals if the drug is efficient in controlling the pest. An example of the perhaps rather unexpected side effect which can occur as a consequence of pesticide treatment in animals, is the sudden dislodging of *Ascaris suum*, the pig roundworm, which may result in a blockage of the gut; in other cases, the disintegration products of parasites, particularly if they are blood dwellers, may cause harmful effects such as embolism or toxaemia.

Phytotoxicity is another important side effect (see Chapters 3 and 4), for example some BHC preparations are toxic to brassicas, spinach, beetroot and potatoes, causing chlorosis of the leaves. The germination of beans can be affected by lindane through its inhibitory effect on mitosis. DDT, although less phytotoxic than BHC, can produce chlorosis in strawberry, tomato and cucumber. Similarly chlordane and some of the organophosphorus pesticides are known to produce phytotoxic effects, and so also do many of the nematicides used against eelworms. Stimulatory effects on germination are produced by sub-toxic doses of lindane, dieldrin and aldrin seed dressings on a variety of crops.

The removal of a pest can also have startling effects on the surrounding flora. After the myxomatosis epidemic of 1954–55 the flora altered markedly in areas previously heavily infested with rabbits. Eroded pastureland became grass covered, the value of marginal land increased, natural regeneration was noticed in many woodlands while in other areas an increase in heather, sedge and gorse took place, all due to the release in grazing pressure of rabbits.

8.1.4 Side effects of pesticides on wild life

In an attempt to control a pest, useful or at least non-harmful animals may also be affected. The accidental destruction of a useful predator may result in increases among undesirable species (Chapter 7). The spraying of crops can have deleterious effects on useful pollinating insects, particular care being necessary where honey bees occur. It is therefore important to prevent spray-drift and to use narrow-spectrum pesticides wherever possible.

It was not until 1956 that the organochlorine pesticides were suspected of being injurious to wild birds and mammals. In 1960, the Royal Society for the Protection of Birds, and the British Trust for Ornithology, published a report on the incidence of suspected pesticide poisoning in birds. As a result investigations were undertaken by the Infestation Control Laboratories of the Ministry of Agriculture, Fisheries and Food, which showed that dieldrin, aldrin and heptachlor used as seed dressing, particularly for spring-sown cereals, were responsible for large numbers of deaths among seed-eating birds and for some resultant poisoning of mammals.

In the summer of 1961 it was therefore voluntarily agreed to discontinue the use of these pesticides for the dressing of spring-sown wheat and to restrict their use for winter sown wheat to those areas of Britain where wheat bulb fly (*Leptohylemyia coarctata*) is prevalent. Although this was a purely voluntary agreement, negotiated between industry and government scientists, it has been strictly adhered to, and few subsequent incidents of poisoning in seed-eating birds have been evident. The significance of residues found in the eggs and bodies of birds is difficult to assess because the rates at which residues build up and break down is not known. Whilst it has also been shown that the organochlorine pesticides in high levels can affect fertility and egg production, these effects are not seen below certain levels of pesticide intake. Consequently the results obtained from chemical analysis must be treated with caution.

The situation with regard to fish-eating and predatory birds does, however, give cause for concern. This is because of the position they occupy at the end of food chains, which renders them vulnerable to the accumulation of large quantities of pesticides from their prey. The 1964 advisory committee on toxic hazards concluded that the use of certain of the persistent pesticides was a major factor in the decline of such British birds as the kestrel, the sparrowhawk and the peregrine falcon. There is little evidence to suggest that the populations of other non-predatory species have been affected, but there is clearly a case to prevent unnecessarily widespread application of highly persistent pesticides.

8.1.5 Precautionary measures

Between the years 1946–50 seven agricultural workers in Britain died following the use of DNOC as a selective herbicide on cereal crops. All these accidents occurred to men actually engaged in spraying. In view of the increasing concern regarding the use of toxic chemicals in Britain the government set up a working party in 1951 to investigate the risks to workers from the use of toxic chemicals in agriculture. Its recommendations resulted in the Agriculture (Poisonous Substances) Act of 1952, which provides for the protection of workers against risks from the toxic chemicals used in agriculture. The Act classifies toxic chemicals according to the degree of danger they may cause, into Part 1 (most dangerous) to Part 4 compounds and also lists the type of protective clothing which must be worn for each class of these compounds. It also specifies the operations in which scheduled compounds can be used. Employers are obliged to provide appropriate protective clothing and to otherwise ensure safe working conditions, and the regulations are rigidly enforced by inspection. It is also obviously important to apply the pesticide only when weather conditions are ideal and when there is the least likelihood of causing harmful effects to non-pest organisms.

Further working parties: *Residues in Food* (1953) and *Risks to Wild Life* (1955) resulted in the setting up of an advisory committee on pesticides and other chemicals which in turn resulted in the formation, in 1957, of a voluntary system of controlling the use of such chemicals. This scheme, formerly known as the Notification of Pesticides Scheme, now called the Pesticides Safety Precautions Scheme, was drawn up between representatives of industry and government. It was revised in 1962–63 and again in 1964. A similar scheme for veterinary products was also introduced in 1964. The purpose of these schemes is to protect man, livestock, domestic animals and wild life against toxic hazards relating to pesticide use. In 1967 the advisory committee set up to review these safety precautions recommended that: "the pesticides and veterinary products Safety Precautions Schemes, should be replaced by a compulsory licensing scheme for all pesticide products used in agriculture, home gardens and food storage and for all agricultural veterinary products except those suitably controlled by other legislation". It also recommends that legislation should be introduced to control the use of certain pesticides and that the possibility of introducing pesticide residue limits for home produced and imported foodstuffs should be considered when sufficient evidence is available. In this context the committee on Food Additives and Contaminants (1967) recommended that statutory limits should be introduced for residues of aldrin and dieldrin in food in the United Kingdom; 0·1 ppm in most food, the exceptions being 1·0 ppm in mutton, 0·003 ppm in liquid milk and 0·02 ppm in baby foods including dried milk.

A further scheme, the Agricultural Chemicals Approvals Scheme, was initiated in 1960 to advise manufacturers on the efficiency and safe use of pesticides. Only pesticides already approved by the Safety Precaution Scheme are considered and if approved the label on the container carries the stamp of approval, a capital A surmounted by a crown.

As a result of the advisory committee's investigations on the persistent organochlorine pesticides in 1961 and 1964 the following recommendations for their safe use were introduced:

1 *The use of aldrin and dieldrin in fertiliser mixtures should cease as soon as this can be arranged.*

2 *The use of aldrin and dieldrin in dips and sprays should cease.*

3 (a) *Seed dressings containing aldrin, dieldrin and heptachlor may continue to be used only on*

  (i) *Winter sown wheat (up till the end of December) only where there is a real danger from wheat bulb fly.*

  (ii) *On rubbed and graded sugar beet seed for precision drilling.*

(b) *Aldrin and dieldrin may be available for commercial use only*

  (i) *Against wireworms in potatoes.*

  (ii) *To control cabbage root fly.*

  (iii) *To control narcissus bulb fly.*

(c) *Dieldrin may be available to control cockroaches and tropical species of ants.*

All other uses of aldrin and dieldrin should cease. It was further suggested that the position should be reviewed in three years and a further report issued.

While considerable adverse publicity has been given to the use of pesticides in Britain, little mention is ever given to the precautions taken. Care is now being exercised, as indicated by the regulations outlined above, to ensure that unnecessary accumulation of pesticides is prevented. While no level of residues, however small, is desirable the risks must be balanced against the undoubted advantages pesticides confer on the community. The emphasis must surely be placed on the search for more effective, specific and less persistent pesticides and the intelligent use of the already existing ones, rather than on the prohibition of a group of chemicals whose undoubted advantages to world food production are already proven. In this context the organophosphorus pesticides, which are highly effective yet not so persistent as the chlorinated hydrocarbon pesticides, are now being used extensively. Further research into the problems of the concentration, distribution, and persistence of pesticides should be encouraged to ensure that the present precautions are adequate.

8.2 Side Effects on the Pest Organism

8.2.1 Resistance

Over a century ago Charles Darwin drew attention to the significance of the small differences that can be observed between the individual animals of a species. This 'variation', which may be structural, physiological or behavioural, occupies a key position in Darwin's Theory of Evolution. Darwin noted that the number of offspring produced by animals far exceeded the number of adults in the next generation and he postulated a 'struggle for existence' among these offspring, of which only the fittest survive. He argued that the 'fittest' are those whose variation equips them to deal more successfully than their fellows with the demands and stresses of life. Superiority in form and function means survival, and survival means the increased opportunity to breed. Darwin assumed the existence of a mechanism for transmitting advantageous features from parent to offspring. Thus each generation was 'naturally selected' by the environment for those variations best suited to meet the strict requirements for survival.

When control measures are adopted against a pest population, man is frequently unconsciously performing an experimental demonstration of the evolutionary process at work. At that moment, the most severe environmental stress on the population is the control measure being applied. The resulting mortality may be very high, and if any pests survive, one is tempted to ask if they are in any sense the 'fittest' members of the original population as far as dealing with this particular 'selective agent' is concerned. An answer is provided by the experience gained in recent decades, which shows that, in many cases, the survivors of control operations are indeed the 'fittest' in the Darwinian sense, and do transmit this 'fitness' to succeeding generations. The individuals possessing a means of reducing the adverse effects of the measures adopted are those most likely to survive, breed, and hence pass on this ability to future populations. This particular evolutionary process leads to what is termed resistance in pests.

The term 'resistance' has already been used in section 2.3 to describe the innate mechanisms by which hosts oppose the threat of pest attack. Now we are using the term to describe the mechanisms by which the pests respond to the influence of control measures. The word is commonly used in both senses and in fact it should be realised that these are both examples of essentially the same process, namely the evolutionary adjustment of a species to counter an adverse influence in its environment.

The most striking examples of resistance developing within pest species result from the use of toxic chemicals as control agents. It was first observed in 1908 when an American orchard pest, the San José scale insect (*Quadraspidiotus perniciosus*) was shown to survive previously lethal sprays of lime sulphur. Four years later, the related citrus pest, California red scale (*Aonidiella aurantii*) was found to have acquired the ability to withstand the highly toxic gas hydrogen cyanide. During the succeeding fifty years the number of examples of resistance to insecticides has grown, this growth being particularly rapid since 1945, when the post-war expansion of the available range of insecticides began. At the present time between 150 and 200 insect species show resistance to at least one insecticide.

The problem is not confined to insecticide application, although the rapidity with which many insects complete their life cycle does predispose them to a correspondingly more rapid appearance of resistance than in the case of slower breeding animals. Exactly the same argument can be applied to Protozoa, and it is not surprising to find that resistance has developed to various anti-protozoal agents both in the case of the African pathogenic trypanosomes and that of *Plasmodium* spp. In theory, of course, any organism can ultimately make the necessary genetic adjustment which would produce resistance and this phenomenon can now be observed in such diverse groups as the bacteria, the parasitic worms and the mammals. In the case of bacteria there are several examples of the development of resistance to drugs, particularly antibiotics, although these problems do not lie within the province of this book. More relevant is the reported development of resistance to the anthelmintic phenothiazine by strains of the large stomach worm of ruminants *Haemonchus contortus* and undoubtedly other examples of anthelmintic resistance are likely to be reported in the future. In the case of mammals, an example of resistance is found in the brown rat *Rattus norvegicus*, which has developed strains tolerant to the widely used rodenticide warfarin. Survival of rats fed on warfarin was first reported on farms in mid-Wales. In spite of attempts to contain this population, resistant rats have subsequently bred in the border counties, and other populations have been reported from the west of Scotland, Gloucestershire and Nottinghamshire. This phenomenon has also now been reported in mice in several parts of Great Britain.

Furthermore, the resistance problem is not wholly confined to the use of chemicals for the reduction of pest populations. In the late 1940s and early 1950s it was shown that some strains of the sawfly pest of larch trees, *Lygaeonematus erichsonii* had become resistant, in North America, to the parasitic wasp *Mesoleius tenthredinis*, which had previously been a useful biological control agent. The parasite egg is encapsulated by phagocytic cells present in the host and these inhibit enbryonic development (section 6.1.4). If this solitary example of a previously valuable parasite being rendered harmless as a result of an evolutionary change in the pest were to be repeated in other instances of successful biological control, using parasites or pathogens, the results might be very serious. In fact, there are already suggestions of increased resistance in flies to a bacillus infection, and in cabbage white caterpillars (*Pieris brassicae*) to virus disease.

Since Darwin put forward his ideas on evolutionary change, the mechanisms by which variation is transmitted from one generation to the next have become far more clearly understood. The heredity factors, or genes, are generally carried on the chromosomes of the gametes, and in the case of genuine resistance one must postulate a gene, or genes, conferring the ability to cope with the particular control measure used.

It should however be mentioned in passing that there are examples, particularly among the higher animals, in which pests acquire what might be loosely termed resistance to control measures during the course of their lives, and this type of resistance is not inherited. Thus, pigeons may become extremely wary of the approach of a man carrying a gun, whereas an unarmed man evokes no flight reaction. Rats and other mammals, may ultimately avoid baits or traps which have been used continuously. This type of 'resistance' is due to the powers of learning possessed by these animals. It is not inherited usually, but must be newly acquired by experience in each generation.

The term resistance is usually confined to the inherited ability of a pest to counteract the control measures taken against it. There has been much discussion as to whether these control measures act in a purely passive way, selecting the resistant genes already present in the population, or whether they actually bring about mutations in the genes which enable the pest, if it survives, to produce resistant offspring. If the first explanation is true, then obviously the genes conferring resistance are already present in the population and the pest is said to be 'pre-adapted' to resistance. The control measure merely sorts out the resistant gene, or genes, concentrating them in subsequent generations. If the second explanation is true, the chemical (in the case of a pesticide application) acts as a 'mutagenic' agent, changing normal genes to ones conferring resistance in the next generation.

In the case of resistance to antimalarial drugs, which has developed in protozoa of the genus *Plasmodium*, it appears that these drugs can act as mutagenic agents. However, insect, mite and rodent resistance to pesticides appears to be pre-adaptive in almost all cases, the chemicals merely selecting the genes which already exist in the population. Almost all experiments designed to show mutagenic effects with sublethal doses of insecticides have failed, but in a very few (one example being the use of dieldrin on certain housefly larvae) mutagenic effects have been suspected.

Usually therefore, resistance will only arise if the population possesses genes capable of producing the required effect. When field populations of a species fail to show resistance after many years of regular treatment, the absence of suitable genetic material is one of the probable reasons. Even if suitable genes are available their selection could also bring together many characters which are in other ways so disadvantageous to the animal that its pest status, or even its ultimate survival is threatened. Nevertheless, pre-adaptation and successful selection occur sufficiently frequently to merit very serious consideration in any discussion of control developments.

There are a number of distinct mechanisms by which pest populations achieve some measure of resistance to pesticides. Animals may be selected for a particular behaviour pattern which enables them either to avoid, or to reduce the time spent in contact with the poison. Thus when a contact insecticide is incorporated into the medium on which the young stages of the fruit fly *Drosophila melanogaster* are reared, the insects developing most rapidly may fail to take up a lethal dose, whereas those developing more slowly are killed. One can in fact produce 'resistant' strains merely by breeding only from the flies emerging from the first pupae formed in a completely non-toxic culture and in this way a 'resistant' population is obtained without any prior exposure to the insecticide, the flies having been primarily selected for speed of development. Similar occurrences take place in nature, an example being certain South American mosquitoes of the genus *Anopheles* which are now unharmed by DDT, simply through avoiding the indoor resting places which are regularly sprayed. Selection with the chemical has produced a population with a new behavioural pattern and the mosquitoes fail to contact the lethal chemical. This is sometimes referred to as 'pseudo-resistance', since a captive animal, if treated with the chemical, is still as susceptible as were its ancestors. 'Pseudo-resistance' also occurs amongst agricultural pests, an example being the tolerance of some codling moth larvae to lead arsenate deposits on apples. This stomach poison is effective if ingested by the larva feeding on the apple's treated surface. The burrowing behaviour of larvae underwent a change in some parts of the world, less of the sprayed surface was consumed, and the deposit ceased to have a lethal effect.

If animals do come into prolonged contact with pesticides, these chemicals must effectively penetrate the body to cause death. Many of the differences observed between the responses of one pest species and another to the same chemical are due to differences in uptake. Species differ in the rate of absorption of stomach poisons from the gut, and in the thickness of the external cuticle through which contact poisons must pass. It has often been suggested that differences between resistant and non-resistant individuals within one species might be similarly due to differences in the barriers presented to the passage of poison into the body. For example, it was thought that resistant strains of scale insect survived fumigation with hydrogen cyanide solely because of a prolonged closure of the spiracles which denied access of external gases to the respiratory system. However, this is not the complete explanation and it is now realised that in resistant pests generally, barriers to penetration are not of major importance.

In some cases, resistance appears to be due mainly to an increased capacity to store the toxic chemical in body tissues where it will cause little harm. Usually however, resistance must be explained in terms of biochemical reactions within the pest. Frequently a detoxification mechanism alters the molecular structure of the poison to a harmless form. A good example is provided by one of the many investigations made into detoxification mechanisms in insects, that involved in the resistance of certain strains of the house-fly, *Musca domestica* to DDT.

Resistant flies contain substantial amounts of an enzyme which destroys DDT by catalysing the removal of hydrogen chloride from the molecule. This enzyme has therefore been called DDT-dehydrochlorinase.

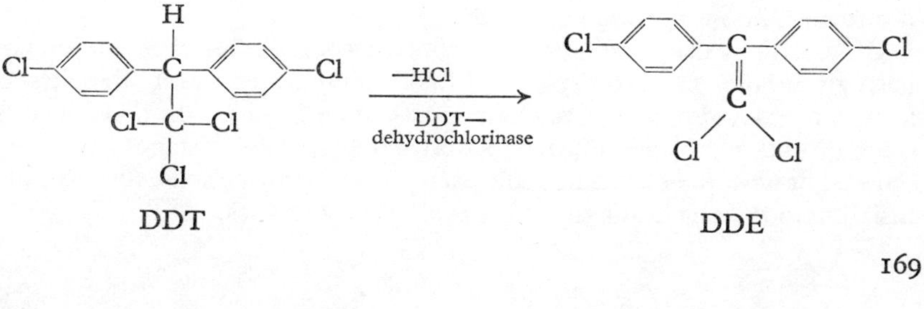

DDT DDE

The substance produced by the reaction is 2, 2-bis-(parachlorophenyl)-1, 1-dichloroethylene, which is universally abbreviated to DDE. The ability to produce large quantities of DDE is found in all resistant strains of houseflies, and their level of resistance is proportional to the amount of DDT-dehydrochlorinase that they contain. The evolution of resistance has not involved the appearance of a new metabolic process, since even normal, susceptible flies contain the enzyme and can convert DDT to DDE. Although they are therefore preadapted to become resistant, the rate at which normal flies carry out this conversion is too slow to save their lives after dosage with DDT. Selection has produced populations containing individuals capable of synthesising abnormally large amounts of the enzyme. Other parallel detoxification mechanisms may, in addition, be present in resistant flies, and there is also some evidence of an enhanced ability to store DDT harmlessly in certain of the body tissues. However, the DDT → DDE conversion is of primary importance to the resistant strains. The same mechanism is found in cases of DDT resistance in other insect species. Some insect pests have never adequately been controlled by DDT, and in a number of these, all members of the species normally contain large quantities of DDT-dehydrochlorinase. These quantities have not evolved as a result of the use of DDT, and the enzyme must presumably serve some other function in the normal life of the insect. Several insect species have evolved DDT resistance which is not dependent upon the DDT → DDE conversion, various other harmless metabolites being produced by quite distinct enzymes.

The relationship between the genetic and biochemical mechanisms underlying the development of resistance is well illustrated by studies on organophosphate resistance in various strains of houseflies. The susceptible flies contain relatively high concentrations of an enzyme termed ali-esterase. The function of this enzyme is obscure, but it does not seem to be of vital metabolic importance. In resistant flies the concentration of ali-esterase is low, but they contain quantities of a new enzyme (a phosphatase) capable of degrading the organophosphate poisons. Genetic studies have shown the inheritance of ali-esterase in susceptible flies, and of phosphatase in resistant flies, to be due to a factor present on the same locus of the same chromosome in each case. It appears, therefore, that the gene controlling the production of ali-esterase in susceptible flies has undergone a mutational change, resulting in the production of a slightly different enzyme molecule. This can no longer function as an effective ali-esterase, but has the new ability to act as a phosphatase, destroying the introduced insecticide.

Of necessity a brief and greatly simplified account of this phenomenon has been given here. The actual picture is more complicated and if other strains are examined additional genes often appear to be involved. In other species quite distinct mechanisms conferring organophosphate resistance have been found, but nevertheless the housefly example does provide some fundamental insight into the practical basis of the evolutionary process.

If treatment of a species with a particular chemical results in resistance developing, other chemicals frequently also lose their effect on the pest. Usually this applies only to chemically related molecules, for example DDT-dehydrochlorinase can also detoxify DDD and methoxychlor, DDT resistant flies being equally unaffected by these other two insecticides. This type of development is known as 'cross-resistance' and normally applies only to chemically related molecules. However, pests which are resistant to one particular chemical group are able to evolve new resistance mechanisms in response to subsequent treatment with new groups of chemicals. These pests are said to be 'multiresistant', and many public health insect pests, which have very frequent contact with insecticides, provide examples of multiresistance problems.

Multiresistance is a particular problem when efforts are made to devise countermeasures against resistance. Thus, the introduction of new compounds into a control programme often produces only a temporary improvement in control. Furthermore, the use of mixtures instead of single pesticides has not always produced satisfactory long-term control, resistance to mixtures having developed, in some cases, almost as rapidly as resistance to single chemicals. Some success is achieved by using 'synergists', which are chemicals which enhance the activity of a pesticide. An example is the substance known as WARF-antiresistant. This chemical inhibits DDT-dehydrochlorinase, and enables DDT to take effect without suffering extensive detoxification. Ultimately, however, several insects have managed to develop resistance to the synergist–insecticide mixture. In particular cases, various other factors can be exploited to counteract resistance. In some species for example, resistance to a particular chemical is correlated with a greatly increased susceptibility to some other chemical. However, as with so many control problems, prevention is probably better than cure. If populations were regularly checked for resistance levels, the establishment of highly resistant forms could perhaps be avoided. This might require frequent switching from one chemical to another, or to alternative control methods, with the aim of maintaining a substantial reservoir of susceptible genotypes within the population.

8.2.2 Other side effects

The development of resistance is one of the major non-lethal results of the exposure of an animal population to pesticides. There are, however, various other ways in which pest animals may be affected by dosage with chemicals, and some of these may have an adverse influence on the pest status of the animal from man's point of view. In some insects, for example, one of the symptoms of poisoning by various insecticides is an accelerated rate of oviposition. This is presumably the result of some neuro-endocrine dislocation and assuming that viable eggs are laid, it is not difficult to imagine a situation in which an inadequate application of insecticide might result in an ultimate increase in the number of offspring of the insect pest.

Good illustrations of these unforeseen influences on pest status are provided by work carried out on various plant sucking insects, such as aphids, which are important vectors of virus diseases of plants. It has been shown, for example, that the wavelength of the light reflected from a leaf surface may be altered by the deposition of a pesticide spray. This can affect the visual attraction exerted by the leaf on the aphid, and experiments with coloured solutions have shown that sprayed surfaces can sometimes be more attractive than unsprayed surfaces. Under these circumstances, therefore, the pesticide application can affect the distribution of the pest population, resulting in greater numbers accumulating on sprayed plants. This might become a particular problem where the spray used is absolutelv non-toxic to the insects, as for example in the case of a fungicide application.

The arrival of an aphid on a plant surface is followed by a sequence of probing activity during which the needle-like mouthparts, or stylets, are thrust into the plant tissues. If the plant is a suitable host, the aphid may remain with its stylets embedded in the tissues, imbibing the sap on which it feeds. If the plant proves unsuitable, probing is followed quickly by withdrawal of the stylets, and the aphid, after an unsuccessful sequence of probes, may move elsewhere. If the aphid probes a virus infected plant, moves to an uninfected plant, and probes again, the disease is likely to be transmitted from one host to the other. Any factor which increases the amount of probing and movement is likely to increase the spread of the disease. Low concentrations of systemic insecticides do, in some cases, appear to repel aphids, resulting in an increased rate of probing and withdrawal. This may explain reports which show that treatment of crops with insecticides has occasionally resulted in an increased incidence of virus disease.

A very similar situation may arise as a result of anthelmintic application to parasitised domestic animals. Some anthelmintics act preferentially on adult worms rather than the young stages. In a number of cases it has been shown that the presence of adult worms inhibits the development of any young worms which subsequently invade, and these then lie dormant within the host. Thus, the use of a non-larvicidal anthelmintic in these cases might have unexpected results, and in, for example, infections by the stomach worm of cattle, *Haemonchus placei*, it has been reported that dosing the host with one of these selective anthelmintics resulted in the expulsion of the adult worms, thus removing the factor inhibiting the ultimate development of far greater numbers of larvae into adults.

These examples of non-lethal effects are of minor economic importance in comparison with the development of resistance. They do illustrate, however, the importance of gathering data on all the biological consequences of a control procedure, if that procedure is to undergo any improvement at a future date.

Chapter Nine
SOME FACTORS ASSOCIATED WITH THE INITIATION, CHOICE AND DEVELOPMENT OF A CONTROL PROCEDURE

9.1 Social and Economic Considerations

The control of injurious animals can be seen to involve considerable technical problems and the expenditure of often considerable sums of money. The extent of financial involvement and trouble to which a person or community is prepared to be committed depends upon their personal, social and economic scale of values, as well as the ability and financial resources to make control possible. Clearly, different standards are applied to the alleviation and prevention of human suffering compared with that of domestic animals. The control of human disease is treated as an ethical problem in most societies but it can only be accomplished within the economic framework of local, national or international policy. On the other hand the desirability and feasibility of non-medical pest control is assessed more strictly in economic terms, since all but the most primitive societies live by trading one with another where prices and profits, and hence purchasing power, are geared to the laws of supply and demand. Clearly, no enterprise rationally sets out to produce more than it can utilise or sell. The utilisation of pest control techniques is just one of many facets of agriculture which helps the formulation of production plans. These are designed to guard the producer against economic and hence social loss, and the producer and consumer against price fluctuations.

The failure of any agricultural product will create scarcity and this can often be filled by more successful competitors elsewhere who will achieve the financial rewards of scarcity value. The absence of economic diversification means that the economies of most of the less well developed countries of the world depend very heavily upon their agricultural production, and its failure is therefore a national calamity, which may further restrict development. Its buying capacity is reduced as are its markets in world trade. Improved pest control technology is an important step towards improving both the output of a nation's labour force and its agricultural production generally, and towards ultimately achieving a reduction of production costs with increased profits. Indeed, the control of such pests of humans as trypanosomes, malarial parasites and schistosomes, and various pathogens of domestic animals within the last decade, has been responsible for the economic and social emergence of certain hitherto backward nations. These nations are now more and more able to contribute to the commerce and livelihood of mankind, and more able to concern themselves with the improvement of stock and crops.

The control of pests and the diseases they cause involves considerable administrative planning, the deployment of specialist research workers and advisory officers, medical and veterinary personnel and equipment, all of which has to be paid for. Despite the very considerable advances in scientific knowledge and technical achievement in this field, and the expenditure of effort and money in developing and administering control procedures, there remains the enormous educational and social problem of making people aware of even the necessity of accepting these measures. Ignorance very often hides behind the veil of religion, social custom, prejudice, and forms of self interest, and therefore education of the population and provision of advisers is of the greatest importance. Internationally, the F.A.O. (Food and Agriculture Organisation) and the W.H.O. (World Health Organisation) (which are financed by the governments of the world, albeit disproportionately) and a number of the more highly developed countries have done a great deal to help with the dissemination of finance and expertise. The aim of this policy is that advancing economic status will itself create resource and expertise which will enable the underdeveloped nations to ultimately help themselves. Regrettably, the new vigour of advancement so often promised is nullified in certain countries by political strife.

The research, development and production of drugs, chemicals and machinery; the breeding of plant and animal varieties; and the training of personnel are extremely costly and must be borne either by governments, or by commercial companies that are permitted to reap adequate financial reward for their efforts.

The rationalisation of control measures is extremely difficult in both biological and economic terms. All too often, in the noted absence of accurate costings an objective economic assessment is impossible and control follows subjective assessments based upon past experience, impressions and calculated guesswork (see also Chapter 1). Startling figures are occasionally produced which attempt to assess losses due to pest attack but these cannot take into consideration the subclinical and undiagnosed losses which inevitably occur.

The scientist is continually concerned with discovering means of increasing output whilst utilising the same, less or different resources. The economist must rationalise these scientific predictions in economic terms and may find that the optimal financial reward is achieved at some level of production short of the maximum possible. Local, national and international trade and finance, availability and demand, determine the optimal financial return for any commodity, and this will fluctuate accordingly—unless the government agrees to pay a guaranteed price for the product. There is always, therefore, a point of reserve production potential which can be exploited if prices justify it. This also means that control of pests is economically justifiable only when the financial return is favourable but one does not usually find out what the return will be until the product is marketed, often some considerable time after control has been instituted.

Mention has been made in section 1.2 of the usefulness of the cost/potential benefit ratio in pest control rationalisation but is very difficult to assess, being reliable only where rates of pest infestation and financial returns are comparatively steady from one time to another. The cost/benefit ratios are therefore worked out only for certain valuable crops such as apples and even then they tend to apply to the benefit likely to follow control of the heaviest infestations.

In practice it is extremely difficult to assess in economic terms whether control measures should be initiated or which method is likely to afford the optimum result. Undoubtedly control measures will continue to be initiated both when they are, and are not, required for some considerable period in the future. It should be emphasised that pest control is only one of many all too poorly understood aspects of agricultural production.

Pest animals do not usually acknowledge man-defined boundaries and individual farmers and communities may sometimes be confronted with the dilemma of whether it is worthwhile instituting control procedures unless neighbours are prepared to do the same. Rural-dwelling peoples the world over are frequently reluctant to commit themselves to enter into co-operative measures unless it is absolutely clear to them that it will profit them to do so. This is a matter of education.

There are a number of control procedures which cannot be instituted and developed except by governments or co-operating communities. It is difficult to envisage, for example, the biological control of pests, methods of sterilisation, and the enforcement of quarantine regulations being handled in any other way. The great popularity of chemical control measures at the present time may in large measure be due to the fact that these, and cultural methods, are often the only means available to the individual farmer.

9.2 Biological Considerations

9.2.1 Assessment of the need for control

Pests present differing control problems according to whether they are liable to invade an area or territory, or whether they are already present. Firstly one can tackle the problem by preventing the pest from reaching a particular area—quarantine (see p. 25). Secondly, one can attempt to destroy the bridgeheads of newly arrived pests or even of already established pests—

eradication. Thirdly, and more usually, if financial resources, expertise, or some other factor, do not permit either quarantine or eradication, an attempt is made to keep well established pests within bounds. Accordingly, the damage caused by the same species of pest animal may be controlled in one region of the world by quarantine, in other regions by eradication, whilst elsewhere attempts are made to keep it down to tolerable levels. As our knowledge of the biology of pests and the technology associated with their control improves, pests diminish in importance and may ultimately be eradicated, at least from those regions where they previously conflicted with man's interests. Once eradication has been achieved then quarantine measures may subsequently function as the means of permanent control. In these ways a number of pests have been eliminated from certain human, animal and crop populations.

This approach is well seen in the world wide attack upon the malarial parasites (protozoa of the genus *Plasmodium*) of man which are transmitted by numerous species of mosquitoes. The progress of this attack is recorded in the Chronicles of the World Health Organisation, the sponsors of the eradication project. Prior to 1960, deficient planning led to lack of success and in April of that year "the Director General of W.H.O. formulated a specific schedule of twenty-two minimum provisions that must be satisfied before a plan of operations can be given technical approval". This included the submission of very comprehensive programme proposals by interested nations as well as realistic estimates of the overall costs of the programme. Needless to say there were many countries in which the general administration and health services had not reached a level at which malaria eradication could be contemplated as laid down in the April 1960 directions. In July 1960, therefore, the W.H.O. Expert Committee on Malaria proposed a new type of assistance for such 'developing' countries which would help in the preliminary operations of malaria control which are adapted to their socio-economic conditions and general development status.

The specific types of projects for which the W.H.O. lends support fall under six headings.

(a) Malarial eradication programmes
(b) Pre-eradication survey projects
(c) Malaria eradication pilot projects
(d) Pre-eradication programmes (for the 'developing countries' as outlined above)
(e) Malaria eradication training projects
(f) Field research projects

All are concerned with the integration of biological research, and new technological achievement as well as the training of competent officials to administer and operate the schemes. The malaria eradication programme (a) is aimed at ending the transmission of malaria by mosquitoes and eliminating the malarial parasite in its human and reservoir hosts. It is a campaign

THE EFFECT OF THURICIDE TREATMENT ON DAMAGE BY CABBAGE LOOPER, *Trichoplusia ni*
(*Courtesy Dr T. A. Angus and the editor of 'World Review of Pest Control'*)

PLATE 8A Cauliflower without Thuricide treatment.

PLATE 8B Cauliflower with Thuricide treatment.

THE EFFECT OF THURICIDE ON GYPSY MOTH DAMAGE TO HARDWOOD TREES.
(Courtesy The International Minerals & Chemical Corporation, Skokie, Illinois, U.S.A.)

PLATE 9A Trees without Thuricide Treatment.

PLATE 9B Trees with Thuricide treatment.

that is limited in time but sufficiently thorough to ensure that at the end of the campaign there is no resumption of disease transmission. It is composed of four stages—preparation, attack, consolidation and maintenance and each involves the activities of a large, well trained personnel. Figure 21 details the population of areas in various phases of malaria eradication, from December 1962 to December 1966, and justifies the summary of the W.H.O. Expert Committee on Malaria, meeting in September 1966:

"... the achievements justify the aspirations expressed by W.H.O. and by previous W.H.O. Expert Committees on Malaria. Difficulties have arisen, as was to be expected. In many cases, these could have been prevented or the

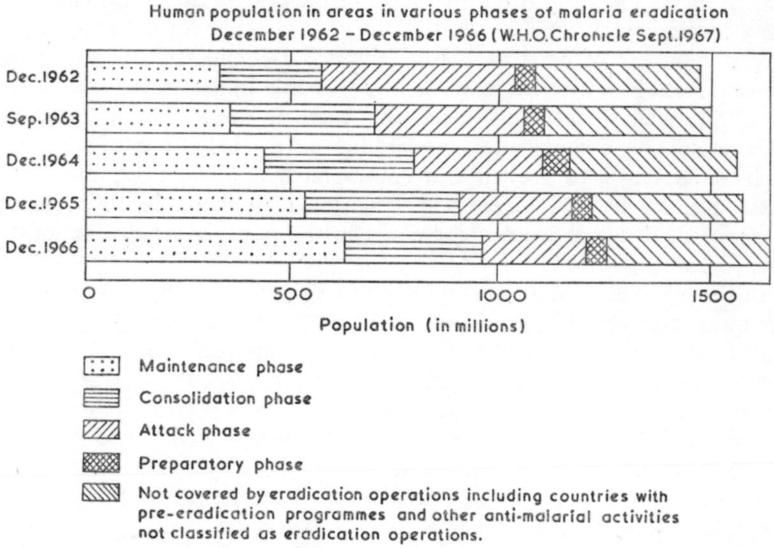

Fig. 21 Human population in areas in various phases of malaria eradication, December 1962–December 1966.

situation corrected by rapid action. Some difficulties have been overcome and the Organisation is making vigorous efforts to solve others. In only eleven years from the first acceptance of the principle, malaria has been eradicated from very large parts of three regions—the European region, the South-East Asian region and the Americas. In two other regions—the Eastern Mediterranean and the Western Pacific—programmes with reasonable prospects of success have been established covering large proportions of the population. In only one region—African—has no progress yet been made . . ."

Of the estimated 1733 million people living in the originally malarial areas of the world (excluding China, North Korea and North Vietnam for which there are no figures), 1366 million (79%) are in areas where malaria has either been eliminated or where eradication programmes are in progress.

In malaria control both the mosquito vectors and the protozoan parasites are attacked. Drug resistance by the parasites and insecticide resistance by the vectors are amongst many biological problems which have to be studied and for which the epidemiologist has to be vigilant. At the end of the attack phase when transmission of malaria has been substantially reduced, then new problems arise. In particular it is important to ascertain the exact level of malaria which remains in the population before deciding whether one can terminate the measures concerned in the attack phase, so that the consolidation phase can ensue. Of particular importance is the part played by the asymptomatic carrier. Even after the eradication programme is completed the problems resulting from migration of infected persons into clean areas are a continuing threat.

Vast numbers of pests cannot at present be eliminated and so we must concern ourselves with keeping them to tolerable limits, taking into account all factors which affect pest population fluctuations and how they interact. Pest population dynamics involves studies on the behaviour of the pest in each of its stages of growth and development and the effect of different environmental factors (in the widest sense) upon these. In this sense the presence of drugs and other chemicals in the pest's environment must be taken into consideration.

In the absence of accurate data regarding the factors affecting population dynamics it can readily be appreciated that forecasting is very difficult and the success that has been achieved is very variable. Ideally, forecasting systems should be as simple as possible for them to be fully workable and intelligible. The liver fluke (*Fasciola hepatica*) forecasting system used in Britain and operated by the Ministry of Agriculture, Fisheries and Food is a simple and working model which incorporates the basic biological knowledge of liver fluke epidemiology but is more particularly reliant upon records of temperature and rainfall at critical times of the year (see Fig. 6).

Forecasting systems require the continuous recording of pest population fluctuations and the advice resulting from this must be capable of indicating the level of attack expected in order to assess whether it will reach the 'economic threshold' whereby practical control measures are worth instigating.

There are reference works which detail methods used for adjudging pest population fluctuations and although the results obtained so far are very variable, there is clearly enormous scope for this complex aspect of control.

9.2.2 *The use of drugs and other chemicals—pesticides*

The problem of developing pesticides against pests of crops shows a number of similarities but also several real differences from that appertaining to pest control in man and other animals, by the very nature of the habitat (i.e. the host itself) in which they are to be used. Any use of a pesticide entails a certain risk, but clearly the amount of risk to which one can subject the human patient is proportionally much lower than it is for domestic animals and crops. Pesticides should be of low toxicity to the crop, animal or man and of as high a toxicity to the pest as possible, and control should preferably be achieved in as few administrations as possible. It should ideally be compatible with other compounds (e.g. other pesticides) and leave no noxious deposit causing taint or otherwise render the crop or animal produce of reduced value. Needless to say, these ideals are comparatively rarely met by existing materials.

In man and other animals, however, where one is dealing with parasites living within the body, often as multiple infections, it is of great advantage if one can use a drug which possesses both a wide spectrum of activity and some degree of persistence. This same desire for wide spectrum insecticides and acaricides for insect and mite pests of crops and animals culminated in the production of such effective pesticides as aldrin and dieldrin, but these not only killed the pest for which they were designed but also large numbers of beneficial insects. Furthermore, they were so persistent that they were carried in the food chains to useful or non-pest animals with subsequent adverse effects (see section 8.1), and their persistence permitted the development of resistant strains of insects. Similarly the persistence of subcurative doses of drugs in animals permits development of drug resistant strains of certain parasites such as trypanosomes (section 8.2.1).

Most insecticides, acaricides, anti-protozoal drugs and anthelmintics vary in their effectiveness against the various stages of a pest's development. Clearly, it is best to use the pesticide at the stage(s) when it can exert its greatest effect, which may not be the time when the pest is doing the most or even any observable damage. Furthermore, one may not be able to use a particularly effective chemical on a crop plant (e.g. apples, pears, etc.) at certain times of the growing season because of the danger of phytotoxicity or of killing pollinating insects. It is also important (see p. 154) to consider factors of natural pest mortality and if possible apply pesticides after these have taken their toll.

Where drug administration to man and animals is contemplated, especially in the control of epidemic or endemic diseases it may be necessary to consider the effects of drug administration upon the acquisition of immunity. The removal of the parasite (or particular stage of the parasite's development) may prevent the host from developing an adequate titre of protective antibody, and hence leave it susceptible to subsequent reinfestation. If drugs are used indiscriminately to remove all stages of parasite development, including those responsible for the development of protective immunity, then a population of animals will emerge which will remain susceptible to reinfection. This may ultimately create difficulties, especially if drug protection is withheld. Animals, and indeed plants, have the power to recover from varying degrees of pest attack, and pesticide usage may merely permit the quicker production of a marketable commodity.

A large number of drugs developed for use in animals and man have a low therapeutic index. An animal that is ill due to parasites is much more susceptible to the toxic effect of drugs than the healthy animal; therefore considerable care should be taken with the use of such drugs. All too often the unqualified farmer will give an overdose of drug 'for luck' with disastrous effects. Underdosing with anti-protozoal agents is also clearly wrong due to their possible mutagenic effect on the parasite (see section 8.2). In order to be treated with antiparasitic drugs for certain disorders, human patients must often enter hospital where an antidote or specialised treatment can be administered if required.

9.3 Commercial Factors Affecting Development of Pesticides

It has already been stated that the expenditure of cash and effort involved in pest control is proportional to that which a population can afford. Scientific and economic research carried out in different parts of the world is published in scientific and other journals and is available to profit others. In this sense, those who have not the basic resource to do the research and development are aided by others who have.

In countries which believe in private enterprise as opposed to state monopoly, the development and promotion of pesticides is largely conducted by commercial undertakings whose interest is spurred by the profit motive. In state controlled undertakings the cost of such work is borne by the taxpayer.

The development of pesticides through all its stages is extremely costly and in order to remain in business and expand in this highly competitive venture any commercial company must keep up its research for products which are commercially viable and which it is possible to use within the strict laws applied by different countries.

Needless to say, any company that hopes to remain in business will have to exploit both the home and the export market for its products. Undoubtedly, many potentially useful pesticides are abandoned during the developmental stages because of their high cost of production which would prohibit their use. Some drugs developed for use in humans are, however, rendered available for humanitarian reasons rather than those of economics.

A company may safeguard its interests by patenting a method or product, if possible, but such patents are of short duration and the company must recoup its outlay over a relatively short period. The further advance of research and technology will ultimately render certain products obsolescent and this in itself creates the necessity for continued scientific and industrial research. In the absence of scientific knowledge which will permit a more rational approach, the search for pesticides remains empirical. The rational approach to pesticide usage through an understanding of the biochemistry of the pest-host relationships and informed prediction of the chemical nature and

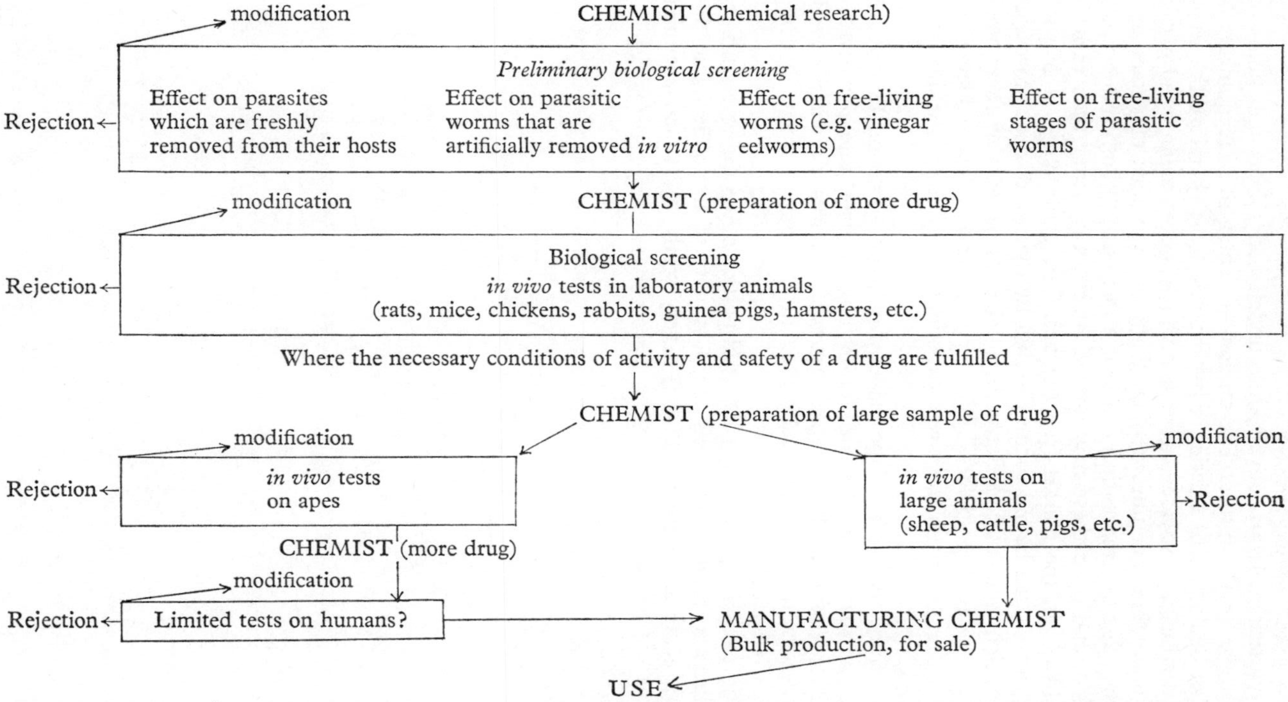

Rejection or modification may occur as a result of drug inactivity, toxic side effects, or for economic reasons.

Fig. 22 Stages in the preparation and screening of parasitic drugs.

activity of pesticides, is a relatively little understood subject and holds no immediate hope for the future. The random screening of large numbers of chemicals, or analogues of existing active chemicals, is therefore the usual method of producing pesticides—a method which is both tedious and costly, and more often than not unrewarding. Thus of thousands of compounds tested by the large pharmaceutical concerns a mere handful of products may be promising, and even these may not be appreciably better than existing products already in the market (see Fig. 22). The enormous cost of such empirical research can be aided considerably by the use of data processing computers which pick out coded structural chemical patterns or compounds contained in the trial pesticides which appear active, thus reducing the chances of missing potentially active compounds by an otherwise subjective assessment.

The empirical approach to pesticide production occasionally has its consolations, in that a drug designed for one purpose may quite unexpectedly find use for another. Thus, certain organophosphorus insecticides, with certain modifications, are successfully used in the control of gut dwelling nematodes of ruminants.

Having developed a potentially successful pesticide it is the responsibility of its producer to examine critically the possibility of any undesirable side effects, both acute and chronic. This is a very costly and time-consuming business.

Chapter Ten
CONCLUSIONS

In this book we have been concerned with the phenomenon of pests and pest control, with the methods and materials currently available for minimising pest damage, and with some of the unforseen resultant side effects. In general, it is clear that there is an ever increasing world-wide demand for the control of pests, and the industry involved in the manufacture and application of pesticides is expanding in an attempt to keep pace with this demand. Since the first pest control company in the United States opened in 1842 in Cincinnati, the industry has grown to the point at which there are now over 6000 such firms in the U.S.A. alone. This overall expansion is well illustrated by the progressive increase in production of pesticides between 1939 and 1963 in the U.S. shown in Fig. 23. Here it will be seen that the consumption pattern for individual pesticides has varied greatly, but the overall increase is clear.

The fundamental reasons for this expansion in pest control lie in the increasing human population, and the increasing standard of living, so that it is probably true to say that the growth in pest control problems is largely a symptom of a deep-seated malaise, the increasing imbalance between man's demands and the availability of food and other material resources.

The chief result of this is that constant pressure is being put upon the pest control expert to get results, and it is not surprising that faced with these urgent problems, he has been forced to respond with swift, but short-term remedies, rather than devote resources to fundamental long-term research, designed to produce permanent solutions. However, it is now increasingly apparent that this heavy reliance on empirically produced poisons must represent a transitional period, as certain serious problems are beginning to develop.

Pesticide resistance

This is a growing problem which is discussed in section 8.2. The development of cross-resistance and multi-resistance has already raised serious control problems in the case of insects and protozoa which frequently come into contact with pesticides. The continual use of one chemical or group of chemicals can be expected ultimately to lead to resistance, and in section 8.2, the need for monitoring the development of resistance has been stressed, so that chemicals can be used in such a way that this difficulty is minimised.

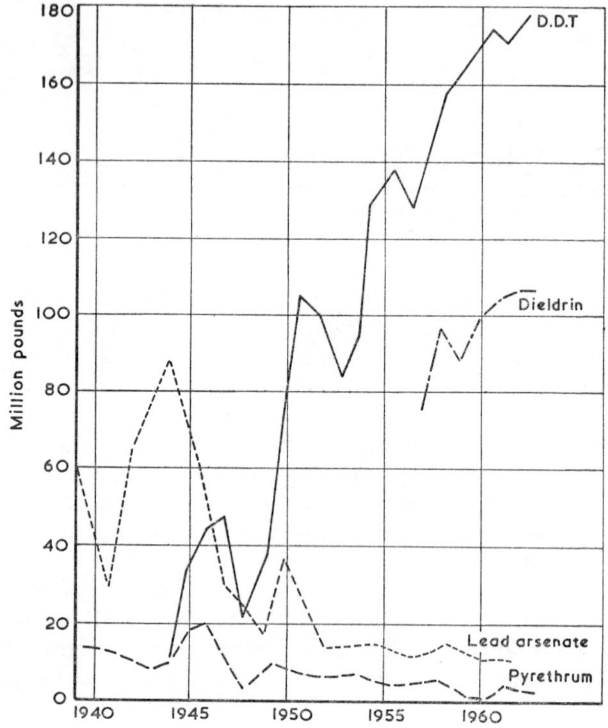

Fig. 23 Changes in the production of insecticides in the U.S.A. 1939–63 (after Moore N. W., 1967).

Nevertheless, the adaptability of living organisms is such that this phenomenon must inevitably impose limits on any control method which depends heavily upon the use of chemicals.

Pesticide addiction

The psychological and ecological pesticide addictions described in Chapter 1 increase our dependence on pesticides if health is to be maintained, or if a worthwhile commodity is to result. This, coupled with resistance and increasing instability in the fauna has made chemical treatments more complex, liberal and costly so that control may be in danger of becoming even more expensive than the original losses due to pests.

Pesticide damage

An over-liberal use of poisons can damage the ecosystem by endangering species other than the pest, including man himself.

One can readily criticise the whole philosophy of pest control by chemical means, and indeed in many quarters it has become fashionable to do so. Nevertheless, one cannot ignore the enormous practical benefits that man has derived, and is deriving from the use of pesticides. The stabilisation of agricultural production has freed many areas of the world from the misery of wasteful gluts of produce followed by the spectre of famine, and in this process the use of pesticides has had no small part to play. However, it is quite clear that we are wielding a two-edged weapon, since the hoped-for successes in the destruction of human disease vectors and parasites is one of the fundamental causes of the population explosion which is creating the ever-increasing demand for food. Although many of the criticisms of the use of pesticides are being met by the development of highly specific chemicals and the enacting of stricter legislation to eliminate residue hazards, it is obvious that a long term solution to pest problems will not come from sole dependence on pesticides. Indeed, pest problems themselves are but a part of a much larger problem of food and commodity production.

It is easy to diagnose that pest control is in a transitional phase; it is much less easy to envisage what possible long-term patterns are likely to result, especially as prediction in science is a notoriously unwise occupation. If current trends continue, however, three possible long-term solutions seem to be emerging.

Firstly, the total exclusion of pests from restricted and controlled environments may provide a solution in certain specialised cases. In 'intensive farming' methods as for example in the production of veal, rabbit, pig meat and poultry products; in food storage and processing; in the maintenance of structures; and in the intensive production of horticultural crops under glass, total pest exclusion is often possible, especially as the environment can be sterilised between each crop. However, it should be remembered that the breakdown of such a system when pathogens gain access can lead to calamitous losses.

Secondly, in the fields of human health, agriculture, forestry and fisheries where it is rarely possible to have the totally controlled pest-free environments discussed above, it may still be possible to eliminate certain types of pest damage. This is best done by protecting the host by immunisation, selective breeding, and the use of repellents or fencing, but in a few cases it may be possible to eradicate the more troublesome pest species. Attempts of this latter sort are fraught with danger, and should only be attempted after the most serious consideration especially as they require large meticulously planned schemes often on a national or international basis. Because of the expense and effort required, it is not surprising that to date, medical pests, particularly disease vectors have been the main target, although the development of sterilisation techniques may well enable this approach to expand.

Thirdly, integrated pest management is inevitably going to be necessary where the total environment cannot be controlled, or where the pest species cannot be eradicated and it is necessary to accept the continued coexistence of pest and host, with the former maintained at tolerable levels of abundance. This in some measure involves a partial return to the more complex and more stable ecosystems found in nature (Chapter 7), and under these circumstances, broad spectrum treatments must be replaced by highly specific chemicals or methods of application. Parasites, predators, competitors, microclimate manipulation and cultural techniques are all utilised and the whole strategy is based on a complex of controlling agents, not on one alone.

Finally, it should be stressed that long-term solutions to pest problems also require human education. Demand, the product of population size and living standards, cannot be allowed to rise indefinitely because if it does, no amount of pest control will be able to guarantee health standards or supply enough commodities for consumption, and moreover, necessity will undermine the whole philosophy of pest management. Eventually, we may have to calculated the yield of each animal or plant crop which we can reasonably expect to get per acre under stable long-term agriculture, and cut the cloth of our demands accordingly; we may also have to re-educate the housewife into accepting a reasonable number of aphids in her cauliflower, and a reasonable number of scars on her apples. Hitherto, pest control experts have not relished the role of re-educators, as it savours of defeat. It may nevertheless be necessary.

In conclusion, it cannot be emphasised too strongly that control problems are not solely the technological problems of the specialist involved in working out the most efficient way of killing an unwanted animal. They are an intimate part of the much larger study of human ecology, i.e., the inter-relationships between man and his living environment, which the complexity of modern civilisation often tends to obscure. Man cannot divorce himself from Nature and at no time in his history has he faced greater threats to the attainment of his basic biological needs for food and survival. To circumvent these threats requires the solution of biological problems, which are the most important and challenging problems with which human beings have ever been confronted.

FURTHER READING

Ascher, K. R. S. (1964) A review of chemosterilants and oviposition inhibitors in insects. *World Review of Pest Control* 3, (1), 7–27.

Brown, A. W. A. (1951) Insect control by chemicals. John Wiley, New York.

Brown, A. W. A. (1960) Mechanisms of resistance against insecticides. *Annual Review of Entomology* 5, 301–326.

van den Bosch, R., and Stern, V. M. (1962) The integration of chemical and biological control of arthropod pests. Ibid. 7, 367–386.

Carson, R. (1962) *Silent Spring*, Hamish Hamilton, London.

Chandler, A. C., and Read, C. P. (1966) *Introduction to Parasitology with Reference to the Parasites of Man*. John Wiley, London.

Crofton, H. D. (1964) Nematode parasite populations in sheep and on pasture. Commonwealth Bureaux of Helminthology. Technical Communication 35.

Crofton, H. D. (1966) *Nematodes*. Hutchinson University Library, London.

DeBach, P. (1964) *The Biological Control of Insect Pests and Weeds*. Chapman and Hall, London.

Deevey, E. S. (1960) The human population. *Scientific American* (Sept.)

Department of Education and Science (1967) Review of the present safety arrangements for the use of toxic chemicals in agriculture and food storage. H.M.S.O., London.

Dethier, V. G. (1956) Repellents. *Annual Review of Entomology* 1, 181–202.

Elton, C. S. (1958) *The Ecology of Invasions by Animals and Plants*. Methuen, London.

van Emden, H. C. (1966) Plant insect relationships and pest control. *World Review of Pest Control* 5, 115–123.

Food and Agricultural Organisation (F.A.O.) (1963) Third World Food Survey. Freedom from Hunger Campaign. Basic Study No. 11. Rome.

Graham, K. (1963) *Concepts of Forest Entomology*. Reinhold, New York.

Geier, P. W. (1966) Management of insect pests. *Annual Review of Entomology* 11, 471–490.

Gibson, T. E. (1964) *The Cost of Animal Parasites*. Span 7, 99–102.

Gibson, T. E. (1969) Advances in Veterinary Anthelmintic Medication. *Advances in Parasitology* 7, 349–373.

Gunther, F. A., and Jepson, L. R. (1960) *Modern Insecticides and World Food Production*. Chapman and Hall, London.

Hickin, N. E. (1963) *The Insect Factor in Wood Decay*. Hutchinson, London.

Jacobson, M. (1966) Chemical insect attractants and repellents. *Annual Review of Entomology* **11**, 403–422.

Jones, F. G. W., and Jones, M. (1964) *Pests of Field Crops*, Edward Arnold, London.

Jones, L. Meyer (Ed.) (1966) *Veterinary Pharmacology and Therapeutics.* Iowa State University Press, Ames. Iowa.

Kilgore, W. W., and Doutt, R. L. (1967) *Pest Control* (*Selected Biological, Physical and Chemical methods*). Academic Press, New York.

Lapage, G. (1963) *Animals Parasitic in Man.* Dover Publications Inc., New York.

Martin, H. (1957) Guide to the chemicals used in crop protection. 3rd Edition, Dominion of Canada, Department of Agriculture Science Survey.

Martin, H. (1964) *The Scientific Principles of Crop Protection.* Edward Arnold, London.

Michel, J. F. (1969) The epidemiology and control of some nematode infections of grazing animals. *Advances in Parasitology* 7, 211–282.

Ministry of Agriculture, Fisheries and Food (M.A.F.F.) Advisory leaflets. Animal Health Leaflets. Technical Bulletins, No. 7 Plant Nematology, 182 Farm Sprays and their Uses. (1955) Toxic chemicals in agriculture—Risks to wild life. H.M.S.O., London. (1961) Toxic chemicals in agriculture and food storage. H.M.S.O., London. (1964) Review of the persistent organo-chlorine pesticides. H.M.S.O., London.

Morris, R. F., and Miller, C. A. (1954) The development of life tables for the spruce budworm. *Canadian Journal of Zoology* **32**, 283–301.

Murton, R. K., and Wright, E. N. (Eds) (1968) The problems of birds as pests. Symposium of the Institute of Biology, U.K. Academic Press, London.

O'Brien, R. D. (1960) *Toxic phosphorus esters.* Academic Press, London.

O'Brien, R. D. (1967) *Insecticides, Action and Metabolism.* Ibid.

Ollerenshaw, C. B. (1959) The ecology of the liver fluke (*Fasciola hepatica*) *Veterinary Record* **71**, 957–965.

Ollerenshaw, C. B. and Smith, L. P. (1969) Meteorological factors and forecasts of helminthic disease. *Advances in Parasitology* 7, 283–323.

Oppenoorth, F. J. (1965) Biochemical genetics of Insecticide resistance. Annual Review of Entomology **10**, 185–206.

Ordish, G. (1952) *Untaken Harvest.* Constable, London.

Painter, R. H. (1951) *Insect Resistance in Crop Plants.* Macmillan, New York.

Papworth, D. S. (1963) How pesticides are approved for safety in Britain. *International Pest Control* 5 (5) 10–13.

P.E.P. (1955) World population and resources. A report by P.E.P. London.

Population Studies No. 41 (1966) *World Population Prospects as Assessed in 1963.* United Nations, New York.

Ripper, W. E. (1955) Application methods for crop protection chemicals. *Annals of Applied Biology* **42**, 288–324.

Rogers, W. P. (1962) *The Nature of Parasitism.* Academic Press, London.

Rose, J. A. (1961) Development of concentrate spraying. *Pest Technology* **3**, 93–96, 134–138, 166–173.

Schnitzer, R. J., and Hawking, F. (Eds) (1963) *Experimental Chemotherapy*, Vol. 1. Academic Press, London.

Schwabe, C. W. (1964) *Veterinary Medicine and Human Health*. The Williams and Wilkins Company, Baltimore, U.S.A.

Smith, C. N., La Breque, G. and Borkovec, A. (1964) *Insect chemosterilants*. Annual Review of Entomology **9**, 269–284.

Smyth, J. D. (1963) *The Biology of Cestode Life Cycles*. Commonwealth Agricultural Bureaux, Farnham Royal, Bucks, England.

Soulsby, E. J. L. (Ed.) (1966) *Biology of Parasites*. Academic Press, New York and London.

Soulsby, E. J. L. (1968) *Helminths, Arthropods and Protozoa of Domesticated Animals* (Mönnig). Bailliere, Tindall and Cassall, London.

Steinhaus, E. A. (Ed) (1963) *Insect Pathology. An Advanced Treatise*. Vol. I and Vol. II. Academic Press, New York and London.

Strickland, A. H. (1965) Pest Control and Productivity in British Agriculture. *Journal of the Royal Society of Arts* **62**, 81.

Sweetman, H. L. (1958) *The Principles of Biological Control*. W. C. Brown Co., Iowa, U.S.A.

Taylor, A. E. R. (Ed) (1968) *Immunity to Animal Parasites*. Sixth Symposium of the British Society for Parasitology. Blackwell, Oxford.

Uvarov, B. P. (1967) Problems of insect ecology in developing countries. Pest Articles and News Summaries, Ministry of Overseas Development, London **13** (3), 202–213.

Watt, K. E. F. (1963) Mathematical population models for five agricultural crop pests. Article in *Population dynamics of agricultural and pest insects*. Editor Leroux *et al*. Memoirs of the Entomological Society of Canada **32**, 83–91.

West, T. F., and Hardy, J. E. (1961) *Chemical Control of Insects*. Chapman and Hall, London.

Worden, A. N., Sellers, J. C., and Tribe, D. E. (1963) *Animal Health, Production and Pasture*. Longmans, London.

Wyniger, R. (1962) Pests of crops in warm climates and their control. *Acta Tropica*, Supplement 7.

Wynne-Edwards, V. C. (1962) *Animal Dispersion in Relation to Social Behaviour*. Oliver and Boyd, Edinburgh.

GLOSSARY

Acanthocephala	Thorny or spiny-headed worms. See under 'helminths'.
Acaricide	A chemical used for the destruction of mites (order Acarina).
Acute	A description of the sudden or rapid response by an animal to a poison or to a disease-causing organism. (As opposed to 'chronic'—see below).
Alternative host	See 'host'.
Amoebic dysentery	Dysentery caused by the protozoan parasite *Entamoeba histolytica* (superclass Sarcodina), when it invades the gut mucosa of man, there typically setting up cradle-shaped bleeding ulcers.
Anopheline	Descriptive of mosquitoes (family Culicidae; order Diptera) which belong to the genus *Anopheles*, the vectors of human malaria.
Anthelmintics	Chemicals used specifically to control infestations of parasitic helminths (see below) in animals.
Antibody	A protein the production of which is elicited by exposure to an antigen (see below) and which combines with that antigen. The proteins are globulins (see below), which are referred to as immunoglobulins. There are many types of antibody, some of which will react with only specific antigens whilst others will react with chemically closely related antigens and are termed non-specific antibodies.
Antigen	A substance which the body recognises as foreign or 'not-self' and which elicits the production of antibody (see above) or hypersensitivity.
Aphids	Insects of the family Aphidae. Plant sucking bugs of the order Hemiptera: Sub-order Homoptera (see below), with a world-wide distribution and numerous species. Common names include 'Greenflies'. Important plant virus vectors.
Arthropod	Literally 'jointed leg'; the Phylum Arthropoda is the taxonomic group containing the animals with articulated jointed limbs and segmented bodies with chitinous exoskeleton, e.g. insects, crustaceans, spiders, scorpions, centipedes, millipedes and mites.

Ascarid Nematodes (see below) belonging to the superfamily Ascaroidea, a group which includes *Ascaris lumbricoides*, the large roundworm of man, *Ascaris suum* the large roundworm of pigs, and *Toxocara* spp., the cat and dog roundworms.

Attenuate To reduce the virulence of a potentially invasive organism; for example by heat, by various types of irradiation, or by treatment with chemicals.

Bark beetles Beetles (order Coleoptera—see below) associated with the weevil group, belonging to the family Scolytidae. The majority of these small insects burrow beneath the bark of trees producing elaborate gallery systems in which the eggs and larvae are reared. Some are serious pests of forest trees and cut timber.

Biomass The total mass of living material of a particular species or community in a given area.

Blowflies Insects of the order Diptera (see below) the larvae (maggots) of which occur in carrion and flesh. Flies of the genera *Calliphora* and *Lucilia* are serious pests in wounds on sheep. Includes blue-bottles and green-bottles.

Boll weevils Beetles (Coleoptera—see below) which feed and breed in the seed vessels of 'bolls' of the cotton plant. The name is commonly applied to the species *Anthonomus grandis*.

Boll worm Also known as the corn earworm, the larva of the moth *Heliothis obsoleta*—order Lepidoptera (see below). It is widely distributed in the world and attacks many economically important crops, particularly tomatoes, maize and cotton.

Booklice Small, usually wingless insects of the order Psocoptera. They are domestic pests which feed on books, papers, cereal products and other vegetable materials. Important 'nuisance-pests' in dwelling houses, libraries and museums.

Bot flies Dipteran (see below) of the genus *Gastrophilus*, the larvae of which are parasitic in the gut of horses and related animals. The adults cause annoyance to the host, flying round it preparatory to laying eggs on its hair. The larvae hatch as a result of licking by the host and are swallowed.

Cabbage Root Fly *Erioischia brassica* a dipteran (see below) which lays its eggs around the stems of brassicas near to soil level. The larvae are extremely destructive to the roots of brassica crops, causing wilting and death in the attacked plants.

Capsids Members of the insect family Capsidae (order Hemiptera, sub-order Heteroptera, see below). These piercing and sucking insects include economically important species attacking fruit crops, tea, potatoes and other agricultural plants on which they act as virus vectors.

Carrot Fly *Psila rosae*, a dipteran (see below) of the family Psilidae. The adult fly lays eggs in the soil and these hatch to produce legless larvae which burrow superficially in the surface of growing carrots.

Cercaria A usually free living (occasionally parasitic) larval stage in the life cycle of many trematode (see below) worms.

Cestodes Tapeworms—see under 'Helminths' below.

Chafers A name applied to certain insects of the coleopteran family (see below) Scarabaeidae. The normally rather swollen C-shaped larvae having reduced legs and live in the soil feeding on decaying vegetable matter, and in the case of the economically important species, on the roots of crop plants. Some of the adult beetles are pests in their own right, since they graze the leaves, flowers or fruit of some commercial crops.

Chemotherapy The process of treatment by the use of chemicals. This term is normally restricted to the use of chemicals against parasites or other disease causing organisms in animals. The drugs and anthelmintics which are used being referred to as chemotherapeutic agents.

Chiggers Mites (see below) of the genus *Trombicula*, the larvae of which infest various reptiles and small mammals but readily attack man. The larvae insert their mouthparts into human skin causing intense irritation and pain. They can act as vectors of 'scrub typhus'.

Chronic A description of a lingering or long-term, and normally sublethal response to a poison or disease causing organisms. (See also 'acute' above.)

Click beetle The adult stage of certain insects of the family Elateridae, order Coleoptera (see below). The name is derived from the noise emitted when the beetle flicks itself into the air after being placed on its back. The larvae of certain genera are the serious agricultural pests known as wire-worms (see below).

Climax An ecological term applied to animal and plant communities denoting a balanced ecological system resulting from the natural succession of organisms, with the eventual domination of some and the subordination or extinction of others.

Clinical An adjective describing the stage in the disease process at which the patient is showing symptoms.

Coccidiosis A disease caused by various genera of the subphylum Sporozoa, phylum Protozoa (see below) in various vertebrates. Examples are found among species of the genera *Eimeria* and *Isospora* which in most cases invade the intestinal epithelium, causing haemorrhage.

Coccids Insects of the family Coccidae. Order Hemiptera—sub order Homoptera (see below) which are commonly known as scale-insects and mealy-bugs. These are very small insects with sucking mouthparts and scaly, waxy or gall-like coverings in which the female is often a highly-specialised and degenerate reproductive sac. Very important pests on a large number of plant species.

Codling Moth *Laspeyrisia (Cydia) pomonella*. Order Lepidoptera (see below). A major pest in all the apple growing areas of the world. The female lays eggs on the surface of the fruitlet or fruit and the larva which hatches burrows down to the core, re-emerging at the time of pupation. This caterpillar is responsible for 'maggoty' apples.

Coleoptera An order of insects in which the fore pair of wings are modified into hard, horny or leathery covers or 'elytra' beneath which the membranous hind wings are folded. These insects are commonly called beetles and numerically, they make up the largest order in the animal kingdom (at least one quarter million species).

Commensalism A regular association between two different species of animals, usually for mutual benefit (at least neither partner being harmed) but where both species are physiologically independent of each other.

Definitive Host See 'Host'.

Diptera An order of insects in which the typical adult has only a single pair of flying wings, the hind pair being reduced to tiny 'balancing' organs. The true flies.

Ecosystem Virtually a more technical term meaning 'Nature'. The animal and plant community interact with their non-living environment to form one ecological system.

Ectoparasite A parasitic organism living superficially on the host (e.g., on skin, fur, or feathers) and not within the body.

Eelworm Small nematode (see below) worms which feed on vegetable matter. The majority of species are harmless soil-dwelling organisms which break down decaying organic matter, but some species attack growing plants and are very serious, economic pests, particularly of cereals, root crops and glasshouse crops. World wide in distribution.

Elephantiasis Disease of human beings caused by filarial worms (see below) of the genus *Wuchereria*. The worms block the lymphatic system producing permanent swollen deformity in the legs or genitalia.

Embolism Any foreign body in the blood stream that is transported from one part of the circulatory system to another, where it becomes lodged in a vessel and causes an obstruction.

Endemic This word is used in two slightly different senses in that a native insect pest, which has not been introduced, is said to be endemic to an area; and a disease which is always found in an area is said to be endemic (contrast 'epidemic' below).

Endoparasite A parasite occurring within the body of the host (contrast 'ectoparasite' above).

Epidemic A strictly relative term applied to the incidence of infection, infestation or disease which appears periodically or occasionally, and clearly in excess of the normal expectancy in the population of an area.

Epidemiology An ecological study concerned with the causes and patterns of infection, infestation or disease.

Fascioliasis Disease due to attack by the liver fluke (*Fasciola hepatica*) in sheep and cattle and many other animals.

Filarial worms Nematode (see below) worms of the superfamily Filaroidea, which inhabit vertebrate blood and other tissues. All require biting fly intermediate hosts (see below), the larvae which these insects pick up from the peripheral blood vessels or tissues of the host being known as 'microfilariae'.

Filariasis Disease caused by filarial worms.

Flea Beetles A large group of small leaf feeding beetles of the family Chrysomelidae. Their common name is due to the fact that they possess swollen hind legs which enable them to jump like fleas. The feeding activity of these beetles riddles the leaves of a number of agricultural crops with holes, and this affects yield and may destroy the crop altogether.

Frit Fly Insects of the dipteran (see above) family Chloropidae. The larvae of these small flies are pests of cereals, particularly oats. Early in the growing season the burrowing larvae destroy the central shoot of the plant, and later generation reduces the seed yield by feeding on the kernels as the grain develops.

Fungicide A chemical used for the destruction of fungi.

Galls Abnormal growth on plants which often results from the presence in the plant tissue of mites, insects (particularly certain members of the Hymenoptera and Diptera) and sometimes eelworms (see above).

Gapeworms *Syngamus trachea* (Nematoda)—the gapeworm of poultry, turkeys, pheasants, etc. Attaches itself to the trachea,

	thus making respiration difficult if present in sufficient numbers. The infested bird typically 'gapes' in an attempt to facilitate respiration.
Globulin	Group of proteins widely distributed in the cells of plants and animals. In animals they are particularly important since they are the main constituents of antibodies—hence the term immunoglobulins.
Grub	Common name for the larval stage of an insect.
Guinea worm	*Dracunulus medinensis* (Nematoda)—a parasite of the subcutaneous tissues of man in warm and arid lands. Infection is achieved by drinking water which contains small crustaceans, the intermediate hosts of the worm.
Helminths	Parasitic worms of the Phyla Platyhelminthes (including the classes Trematoda (flatworms) and Cestoda (tapeworms)), Acanthocephala (thorny-headed worms) and Nematoda (roundworms or nematode worms).
Hemiptera	An order of insects, the members of which are commonly known as 'bugs' and have two pairs of piercing needle-like mouthparts. Many species cause serious damage to crops by their feeding activities, and through the transmission of plant viruses. A few suck the blood of vertebrates and act as disease vectors.
Heteroptera	One of the sub-orders of the Hemiptera (see above) in which the fore wings are hardened as a covering for the hind wings, but retain a very obvious membranous area at the tip.
Homoptera	The other sub-order of the Hemiptera (see above) in which the wings, if present, are uniformly membranous and usually folded in a sloping fashion over the sides of the body. Wingless forms are frequent.
Hookworms	Strongyle (see below) nematodes belonging to the family Ancylostomidae. So-called because of the cutting plates in the mouth with which they lacerate the tissues of the host in order to suck blood. Important intestinal parasites of a large number of mammals including man.
Host	An organism infected or infested by a pest. Alternative host: an organism which may act as a definitive host instead of that which is usual. Definitive host: the host which harbours the adult or sexual stage of a parasite. Intermediate host: the host which harbours the asexual phase(s) of a parasite. Reservoir host: Usually a wild host which harbours the pests of domesticated animals or plants.
Hymenoptera	An order of insects in which the fore and hind wings on either side are coupled together by hooks to form what

is functionally a single membrane. This order includes the wasps, sawflies, bees and ants.

Immunity See page 56

Intercurrent infection Infection by two or more different organisms.

Intermediate host See 'host' above.

In vitro Describing the culture of living material, or the experimentation on living material which is carried out in the test tube, i.e., divorced from the normal living environment.

In vivo The opposite of *in vitro* (above), e.g., an experiment carried out on a parasite contained within its normal host is an *in vivo* experiment.

Leafhoppers The common name for insects belonging to the Hemiptera-Homoptera (see above), family Jassidae. Apart from the aphids, this is the most abundant group of bugs and contains species which are serious pests of agricultural crops and fruit and forest trees all over the world. The name is due to the fact that members of this group typically use their elongated back legs for jumping, in some cases clearing distances of several feet.

Leatherjackets Common name for the larvae of insects of the dipteran (see above) family Tipulidae. The adults (daddy-longlegs or crane-flies) lay eggs in the soil or other damp situations and the larvae which hatch typically feed on roots or decaying vegetable matter. Leatherjackets are a particular pest of turf but also attack arable crops.

Leishmaniasis A disease caused by a member of the protozoan genus *Leishmania*, transmitted by sandflies (*Phlebotomus* spp.—Order Diptera), *L. donovani* causes Kala-azar, *L. tropica* causes tropical sore, and *L. braziliensis* causes espundia.

Lepidoptera An order of insects in which the two pairs of wings and the body are clothed in scales, and the mouthparts typically consist of a suctorial proboscis. Commonly known as the butterflies and moths, the larvae (caterpillars) are often serious pests of agricultural crops, forest, trees, and stored materials.

Lice A common name which usually refers to wingless insects of the order Siphunculata which are ectoparasites (see above) of mammals and use their piercing mouthparts to suck blood. However, birds carry ectoparasites, referred to as feather lice or biting lice (insect order Mallophaga) which feed on or among the feathers. The word lice is also used rather more loosely in that aphids (see above) are sometimes called plant lice, and psyllids (see below) are referred to as jumping plant lice.

Locus The position on the chromosome which a particular gene occupies is referred to as the 'locus' of that gene.

Mange A condition in man and domestic animals resulting from infection of the skin with various species of mites (see below) including *Sarcoptes* spp. causing sarcoptic mange in man, cattle, sheep, pigs and dogs, and *Psoroptes* spp. causing psoroptic mange (sheep scab) in sheep and goats.

Mealy bugs A sub-group of the hemipterous family Coccidae (see above) which are sometimes known as pseudococcids. The adult female insect is a sedentary organism, normally secreting long filaments of protective white wax or 'wool'. Various species are serious pests of fruit and other crops in the tropics, and glasshouse crops in the more temperate regions of the world.

Metamorphosis The insect which hatches from the egg is usually different from the final adult stage. The change from the juvenile to the adult form is referred to as the metamorphosis of the insect and may take place gradually throughout the pre-adult life, or may occur relatively abruptly during a pupal stage.

Metazoa One of the two major subdivisions of animal kingdom (cf. Protozoa—below) in which are grouped all the animals whose bodies consist of more than one cell and which show some degree of organisation of these cells into functionally specialised tissues.

Mite The common name for arthropods of the Acarina. This order belongs to the arthropod (see above) class Arachnida, the members of which differ from the insects in the possession of four (instead of three) pairs of legs, and the absence of antennae. (Apart from mites, the class includes scorpions and spiders.) Mites are small, and very abundant organisms with piercing and sucking mouthparts. Their diverse feeding habits on a great range of plant and animal material have resulted in many species acquiring pest status.

Molluscicide A chemical used for killing molluscs (slugs and snails).

Moniezia A genus of tapeworm (see 'helminths') commonly found in the intestine of sheep and cattle in many areas of the world, and which utilises oribatid mites (see below) as intermediate hosts.

Morbidity Clinical (see above) disease.

Necrosis A limited amount of death of tissue in a living organism.

Nematode The round worms (see 'helminths' above).

Onchocerca volvulus A filarial worm which infests the skin. The worms occur in dermal nodules where their microfilariae are deposited and from which the latter migrate through the tissues of the dermis. Transmitted by flies of the genus *Simulium* which live near fast flowing streams.

Oribatids One of the subdivisions of the Acarina, or mites (see above) which because of their shiny, horny body surface are commonly called 'beetle mites'. Most species live in the soil or decaying vegetation. (See also *Moniezia*—above).

Ovicide A chemical used for destroying ova (eggs), usually of insects or mites.

Paramphistomum A genus of trematodes (see 'helminths' above) belonging to a family characterised by the presence of a large posterior sucker. Members of the genus are cosmopolitan in distribution and occur in the gut of ruminants particularly sheep and cattle.

Pathogen Any disease-producing organism.

Perennating (of plants) Survival from year to year by vegetative means.

Phagocytosis The process by which cells flow round and engulf particles. Phagocytes are cells of importance in the defence of most of the Metazoa (see above), and in the case of man and other mammals they include the polymorphonuclear leucocytes, and macrophages of the blood.

Pinworm or Seatworm *Enterobius vermicularis*, a very common parasite of man, especially of small children. The adult worms live in the large intestine and the female visits the anus to lay her eggs. Her activities and the sticky eggs cause irritation which induces scratching. This activity transfers eggs to hands, clothing, etc., from whence they may be transported to the mouth of non-infected persons.

Piroplasms A group of protozoans (see below) parasitic in the blood system of a wide variety of animals and transmitted by blood sucking ticks (see below). Examples are *Babesia spp.* causing red water fever in cattle.

Prophylaxis The prevention of disease.

Prothrombin The factor in blood plasma which is the precursor of thrombin. Thrombin is an enzyme which converts fibrinogen into fibrin, a process which is responsible for blood clot formation.

Protozoa A sub-kingdom of animals in which all the members are unicellular. Some species aggregate to form multicellular colonies but no single nucleus has control of a specialised portion of the cytoplasm (contrast Metazoa above). The Protozoa includes pests such as parasitic amoebae, malaria causing organisms and trypanosomes.

Pseudococcids See mealy bugs (above).

Psyllids Insects of the order Hemiptera, sub-order Homoptera (see above). These are very active small bugs which leap and fly among plant foliage (hence common name—jumping plant lice). The family to which these insects belong is the Chermidae, or Psyllidae, and species of the genus *Psylla* are very injurious pests of fruit trees.

Pulses Edible seeds of leguminous plants, e.g., peas, beans and lentils.

Reservoir host See 'host' above.

Rodenticide A chemical used for the destruction of rodents (e.g. rats and mice).

Rootfly Fly larvae feeding on plant roots, see 'cabbage rootfly' (above).

Roundworms See 'helminths' (above).

Sawflies A sub group of the Hymenoptera (see above) in which the female is equipped with a saw-like ovipositor with which she cuts a slit in plant tissues in order to insert her eggs. The larvae which hatch resemble lepidopteran (see above) caterpillars in their general appearance and some are serious defoliators of vegetable, fruit and forest crops.

Scale Insects See 'coccids' (above).

Schistosomes Blood dwelling flukes of the genus *Schistosoma*. Various species are found in man, cattle, buffaloes, other domestic animals, wild animals and birds. Very important in man causing very considerable debility in many warmer parts of the world, wherever the snail intermediate hosts are found. The cercariae of these flukes are able to penetrate skin and infection is therefore most often found in man and animals that regularly work in, bathe in, or drink infected water.

Scolex The anterior end or 'head' of typical tapeworms. It is often equipped with hooks and/or suckers with which it is attached to the gut lining of the host.

Serological Pertaining to the study of blood serums and the activity of their constituents, especially antibodies (see above).

Spirochaetes Bacteria-like single celled organisms which are often responsible for producing disease in a variety of animals, syphilis in man being an example.

Strongyles An anglicised term for those nematodes (see 'helminths' above) which belong to the sub-order Strongylata. This sub-order includes many important parasites of domestic animals, for example the hookworms (see above), the redworms of horses, and the gapeworm (see above) of poultry.

Sub-clinical stage A stage in the disease process without clinical (see above) manifestations.

Taeniaecides Chemicals used for the destruction of tapeworms (see 'helminths'—above) of the genus *Taenia.*

Tapeworms Worms of the group Cestoda (see 'helminths'—above).

Tarsi The small terminal 'segments' of the jointed leg of an insect. The number of tarsal joints varies from stage to stage and species to species. The terminal segment may carry hooks or pads to assist the insect in gripping the surface over which it is walking.

Termites Insects of the order Isoptera, which are commonly known as 'white ants' although, completely unrelated to the true ants (see 'Hymenoptera'—above). Termites are colonial, social insects which are best known because of the mound building activities of some species. Alternatively colonies may be established in structural timber, causing very serious damage. Tropical pests attacking various agricultural crops, forest trees and man-made structures.

Therapeutic Pertaining to the treatment of disease, a therapeutic dose of an anthelmintic, for instance, is one of sufficient quantity and quality to fulfill the purpose for which it is employed.

Thorny-headed worms The Acanthocephala (see 'helminths' above).

Thrips Small insects of the order Thysanoptera which possess peculiarly asymmetric piercing and sucking mouthparts. Various species of thrips are serious pests of vegetable, fruit and flower crops, and are incriminated as plant virus vectors.

Ticks A group of arachnids (see under 'mites' above) grouped with the mites in the Acarina. They are large leathery mites which are usually temporary ectoparasites (see above) of mammals, birds and reptiles. The blood sucking activities of ticks enable them to transmit a number of diseases to man (e.g. relapsing fever) and domesticated animals (e.g. piroplasms—see above).

Tipulid A fly of the family Tipulidae (see 'leatherjackets' above).

Titre The volume of a solution required to react with, or correspond to, a given volume of another solution. E.g., in serological work this has come to be used as a means of expressing the amount of antibody produced against homologous antigen.

Tortricids Lepidopterous (see above) insects of the group Tortricidae, the larvae of several species of these moths being serious defoliating pests of fruit and forest trees, and some arable crops.

Toxaemia The presence in the blood of poisonous (toxic) products formed in the body cells. These may be produced as a result of dosing with chemotherapeutic agents, or as the result of invasion by parasitic organisms.

Toxoplasmosis Disease caused by *Toxoplasma gondii*. A cosmopolitan parasite of most vertebrates; a coccidium (Sporozoa) of cats, it has been incriminated in causing still-births in humans, retinal lesions, meningitis, fevers, pneumonia and heart lesions.

Trematode See 'helminth' above.

Trichinosis A disease in humans resulting from invasion of the voluntary muscles of the body by larvae of the nematode (see above) worm *Trichinella spiralis*. The disease is normally contracted by man through eating poorly cooked pork contaminated with encysted larvae.

Trypanosomiasis A disease of man and other vertebrates resulting from the infection of the blood with protozoan (see above) parasites of the genus *Trypanosoma* usually transmitted by tsetse flies and other biting flies.

Vaccination The administration of either attenuated or killed organisms, or antigenically related organisms, to cause antibody formation and produce immunity to infection by the specific organism.

Vector An animal which carries the infective stages of parasites and transmits them to the definitive or final host, e.g., the tsetse fly which is a vector of trypanosomiasis (see above).

Warble Flies Members of the order Diptera (see above) belonging to the genus *Hypoderma*. The larvae of these insects burrow in the tissues of cattle, ultimately locating in the back where they form prominent lumps, or warbles. Emergence of the fully-developed larvae produces holes in the hide which drastically lower its commercial value.

Weevils Beetles (order Coleoptera—see above), of the group Curculionidae. Weevils generally have the face drawn out into a long snout or rostrum. Many are serious pests of plants and stored products in both the larval and adult stage.

Winter Moths Lepidopterans (see above) in which the females are flightless and ascend trees to lay eggs on foliage buds early in the year. The larvae are serious defoliators and may damage fruit occasionally.

Wireworms Larvae of click beetles (see above). There are several species which feed on the roots of plants, doing particularly serious damage to cereal and root crops.

INDEX

(Bold figures refer to those pages on which illustrations occur)